Studies in Infrastructure and Control

The book series aims to publish top-quality state-of-the-art textbooks, research monographs, edited volumes and selected conference proceedings related to infrastructure, innovation, control, and related fields. Additionally, established and emerging applications related to applied areas like smart cities, internet of things, machine learning, artificial intelligence, etc., are developed and utilized in an effort to demonstrate recent innovations in infrastructure and the possible implications of control theory therein. The study also includes areas like transportation infrastructure, building infrastructure management and seismic vibration control, and also spans a gamut of areas from renewable energy infrastructure like solar parks, wind farms, biomass power plants and related technologies, to the associated policies and related innovations and control methodologies involved.

Sombir Kundu · Ashutosh K. Giri ·
Dillip Kumar Mishra
Editors

Hybrid Renewable Power Infrastructure for Sustainable Electric Vehicle Development

Springer

Editors
Sombir Kundu
Department of Electrical Engineering, MMEC
Maharishi Markandeshwar (Deemed to be University)
Ambala, Haryana, India

Ashutosh K. Giri
Department of Electrical Engineering
Government Engineering College
Bharuch, Gujarat, India

Dillip Kumar Mishra
School of Electrical Engineering
Kalinga Institute of Industrial Technology (KIIT) (Deemed to be University)
Bhubaneswar, Odisha, India

ISSN 2730-6453 ISSN 2730-6461 (electronic)
Studies in Infrastructure and Control
ISBN 978-981-95-1322-2 ISBN 978-981-95-1323-9 (eBook)
https://doi.org/10.1007/978-981-95-1323-9

This Springer imprint is published by the registered company Springer Nature Singapore Pte Ltd.
The registered company address is: 152 Beach Road, #21-01/04 Gateway East, Singapore 189721, Singapore

Contents

1 **Introduction of Fuel Cell-Based Hybrid Charging Infrastructure for Electric Vehicles** ... 1
Sombir Kundu, Ashutosh K. Giri, and Sunil Kadiyan

2 **Energy Storage Solutions and Induction Motor Drives for Sustainable EV Charging** ... 17
Vaishali Rai, Arpit Yadav, Dillip Kumar Mishra, Manob Hazarika, Nikunj R. Patel, and D. P. Kothari

3 **Hybrid Charging for Electrical Vehicle** ... 31
Jigneshkumar P. Desai

4 **Machine Learning-Based Grid-Interactive Chargers for Optimizing Power Quality Under Non-grid Conditions** ... 47
Gaurav Yadav, Sudhanshu Mittal, and Sombir Kundu

5 **Integration of Fuel Cell–Renewable Energy Sources in Hybrid Infrastructure** ... 59
Preeti Sharma, Shubham Priyadarshi, Anakshi Boruah, and Rajneesh Kumar

6 **Power Quality Enhancement in Grid-Connected EV Charging Systems Using Third-Order Sinusoidal Integrators to Control BADC** ... 71
Lochan Varshney, Chaudhary Indra Kumar, and Prakash Chittora

7 **Electric and Hydrogen Vehicles: Systems, Infrastructure, and Future Trends** ... 85
Kanchan Bala Rai

8 **Transitioning to Electric Mobility: Challenges, Opportunities, and Future Directions** ... 111
Sunil Kadiyan, Sombir Kundu, Jitender Kumar, and Aman Aggarwal

9 Hybrid EV Charging Systems: A Study Report on Integration of Fuel Cell with EV for Clean Solution 123
Nitin Sundriyal, Sombir Kundu, Sandeep Gupta, Ankit Bhatt, and Navdeep Singh

10 Renewable Energy-Based EV Battery Swapping Stations: Opportunities and Future Directions 141
Prasanta Kumar Mohanty, Rudranarayan Pradhan, Premalata Jena, and Narayana Prasad Padhy

11 Case Study for Integration of Renewable Sources to Loharu Micro Grid to Reduce the Expenses and Emissions Using HOMER Pro 169
Surender Singh

12 Future Trends and Innovation in EV Charging 189
Deepesh Sharma, Mohit Kumar, and Atma Ram

About the Editors

Sombir Kundu 2011, M.Tech. in Electrical Engineering with a specialization in Power System from the DCRUST, Murthal, Sonepat, Haryana, India, in 2014 and Ph.D. from the Department of Electrical Engineering, Delhi Technological University, New Delhi, India, in 2023. In August 2014, he joined the Department of Electrical Engineering, School of Engineering and Technology (a unit of Ganga Technical Campus), Bahadurgarh, Haryana as an assistant professor. In May 2022, he joined the Department of Electrical Engineering, Ganga Institute of Technology and Management, Jhajjar, Haryana, India, as an assistant professor. In August 2024, he joined the Department of Electrical Engineering, MMEC, Maharishi Markandeshwar (Deemed to be University), Mullana, Ambala, Haryana, India, as an assistant professor. He has authored and co-authored more than 10 research papers in reputed journals such as Elsevier, Wiley, Tailor Francis, Springer. He has also authored and co-authored three book chapters which are published in Elsevier Publication and CRC Press. He has served as a session chair for the International Conference on Intelligent Computing, Simulation, and Optimization, December 08–10, Goa.

Ashutosh K. Giri received his Bachelor of Engineering in Electrical Engineering from College of Technology, G. B. Pant University of Agriculture and Technology, Pantnagar Uttarakhand, India, in 2002, M.E. in Electrical Engineering with specialization in Power System from the L. D. College of Engineering, Ahmedabad, Gujarat, India, in 2005 and Ph.D. from the Department of Electrical Engineering, Sardar Vallabh National Institute of Technology, Surat, India in 2020. In May 2011, he joined the Department of Electrical Engineering, Government Engineering College, Bharuch, as an assistant professor. He has authored and co-authored more than 60 research articles in reputed journals such as IEEE Transactions, IET Journal, Wiley, Tailor Francis, Springer. He has also authored and co-authored 10 book chapters published in Elsevier, Springer Publication, and CRC Press. Also, he has edited or authored five books in Elsevier, CRC Press, and Springer Nature Singapore, respectively. He is working on four sponsored research projects. He is supervising or co-supervising six Ph.D. students (one completed in DTU New Delhi and five in progress in GTU, Ahmedabad, and other universities) and coordinated six Short Term

Courses. He has delivered more than 35 national and international level expert talks at various reputed universities across India. He had received a Research Opportunity Week (ROW) scholarship and visited Technical University of Munich Germany in March–April 2019.

Dillip Kumar Mishra (S'15, M'22, SM'24) earned his Ph.D. in Electrical Engineering from the University of Technology Sydney and completed his postdoctoral fellowship at Clemson University. Currently, he serves as an Assistant Professor in the School of Electrical Engineering, KIIT University, India, and prior to that, he was the Research Assistant Professor in Automotive Engineering, Clemson University, USA.

Dr. Mishra is actively involved in IEEE leadership, holding positions as Chair of the IEEE Young Professionals, Piedmont Section (USA), and Vice-Chair of the IEEE Consumer Technology Society (Global). Additionally, he was the President of the Postdoctoral and Early Professional Society (PEPS) at Clemson University.

With over 50 publications, including 19 journal articles, 22 conference papers, 10 book chapters, and two books, Dr. Mishra has made significant contributions to his field. He serves as an editor and guest editor for prestigious journals and publishers such as Wiley, Elsevier, *Frontiers in Energy Research,* and *IEEE Transactions on Consumer Electronics* (Special Section). He has also chaired sessions at the IECON 2025 conference and served as an Associate Editor for the IEEE Control Systems Society Conference.

Bringing a decade of combined experience across industry, academia, and research, Dr. Mishra possesses deep expertise in batteries, electric vehicles (EVs), renewable energy systems, power system resilience, energy markets and microgrids.

Chapter 1
Introduction of Fuel Cell-Based Hybrid Charging Infrastructure for Electric Vehicles

Sombir Kundu, Ashutosh K. Giri, and Sunil Kadiyan

Abstract As the world shifts toward sustainable energy and the adoption of electric vehicles (EVs), the seamless integration of renewable energy sources (RESs), microgrids, and charging infrastructure for EVs becomes crucial. This chapter offers a comprehensive examination of the latest advancements in combining solar and wind energies, fuel cells, and microgrids with electric vehicle charging systems. It explores important topics like energy management strategies, maintaining grid stability, and establishing dependable charging infrastructure. The unpredictability and inconsistency of renewable energy, along with the intricacies of managing and expanding energy systems, are emphasized as major challenges. The evaluation of microgrids focuses on their capacity to enhance energy security and system resilience by incorporating distributed energy resources (DERs), such as fuel cells, and optimizing power generation and consumption. The chapter also evaluates the efficiency of energy storage systems (ESS) and bidirectional vehicle-to-grid (V2G) technologies in reducing peak demand and maintaining energy equilibrium. Technologies like smart grids, advanced power electronics, and communication protocols are examined because they are essential for achieving seamless integration within the system. Furthermore, the chapter delves into significant standards and policy frameworks, as well as future research directions necessary to expedite the development of a reliable, scalable electric vehicle charging infrastructure that integrates microgrids, fuel cells, and RESs. The discoveries offer valuable information for creating sustainable energy strategies that promote the widespread use of EVs and facilitate the shift toward a low-carbon energy future.

S. Kundu (✉)
Electrical Engineering Department, MMEC, Maharishi Markandeshwar (Deemed to be University), Ambala, Haryana, India
e-mail: sskundu46@gmail.com

A. K. Giri
Electrical Engineering Department, Government Engineering College, Bharuch, Gujarat, India

S. Kadiyan
Mechanical Engineering Department, Ganga Institute of Technology and Management, Jhajjar, India

S. Kundu et al. (eds.), *Hybrid Renewable Power Infrastructure for Sustainable Electric Vehicle Development*, Studies in Infrastructure and Control,
https://doi.org/10.1007/978-981-95-1323-9_1

Keywords Electric vehicles · Microgrids · Smart charging · Charging infrastructure

1 Introduction

The increasing popularity of EVs as a cleaner alternative to traditional internal combustion engine vehicles presents both exciting opportunities and complex challenges for the current power infrastructure. Simultaneously, the worldwide transition toward RESs, such as solar and wind, necessitates their seamless integration into the power grid. Constructing a sustainable and resilient energy infrastructure requires the development of advanced microgrid systems that seamlessly incorporate electric vehicle charging infrastructure, renewable energy generation, energy storage, and emerging technologies such as fuel cells [1–3]. As the transportation sector transitions toward electrification, it is crucial to integrate essential elements such as electric vehicle charging networks, fuel cells, and RESs in a synchronized and effective manner. ESS minimizes the inherent uncertainty of energy production by storing surplus energy generated during peak solar hours and providing it during periods of low production or high demand [4, 5]. Similarly, fuel cells provide flexibility and reliability by converting hydrogen into electricity, generating clean and consistent power that can supplement or act as a backup for intermittent RESs. Microgrids, which function as independent and localized energy systems, are crucial in facilitating the integration of DERs, including RESs, energy storage devices, and fuel cells [6, 7]. These advancements enhance energy management capabilities, decrease dependence on the central grid, and contribute to the reduction of greenhouse gas emissions. Nevertheless, the fluctuating nature of RES, including variations in solar irradiance and wind speeds, continues to present significant challenges to grid stability and consistent power delivery [8, 9]. The rapid advancement of EV technology, coupled with the global transition to RESs, is reshaping the current power infrastructure. As the adoption of EVs grows, cutting-edge charging technologies such as v2h and v2g emerge, enabling EVs to serve as mobile energy storage devices that can exchange energy in both directions [10, 11]. These technologies allow EVs to not only draw power from the grid, but also to send back stored electricity to homes or the grid during times of high demand, grid strain, or power outages. V2H and V2G systems allow EVs to act as backup energy sources, reducing dependence on centralized grid infrastructure and enhancing overall energy resilience [12, 13]. These technologies play a crucial role in promoting demand-side flexibility, maintaining grid stability, and facilitating the integration of RESs such as solar and wind by releasing energy during peak demand periods and replenishing during off-peak hours. Nevertheless, to fully harness the benefits of these interactions, sophisticated control algorithms, secure communication protocols, and real-time energy management systems must be integrated to effectively manage the flow of energy between electric vehicles, the power grid, and distributed RESs [14, 15]. In addition to the technological landscape, the legal and policy environment plays a crucial role in determining the successful implementation of integrated renewable energy [16, 17] and EV charging infrastructure [18]. The future of energy innovation and interoperability is formed by key international frameworks, such as the EU Green Deal's commitment to carbon neutrality, V2G ISO 15118, IEEE 2030.5, intelligent energy protocols, and FERC order 841 for fair energy access. These activities are essential for controlling integrated renewable energy systems, ensuring the resistance of the grid, and supporting global sustainability [19, 20].

2 Methodology

This extensive study carefully investigates the integration of microgrids, EV charging infrastructure, RESs, and fuel cells, to synthesize current knowledge, identify important issues, and assess potential solutions. A structured approach was used to conduct a complete literature review, which included peer-reviewed journals, conference proceedings, technical reports, and industry white papers. This analysis made use of key resources such as IEEE Xplore, Science Direct, Springer Link, and Google Scholar [21–23]. The literature research focuses on papers from 2014 to 2025, a decade that saw tremendous progress in the convergence of microgrids, EV technologies [24], renewable energy integration [25, 26], and alternative energy systems such as fuel cells [27]. This timeline was chosen not just for its technological significance, but also because it coincides with significant global milestones such as the 2015 Paris Agreement and the acceleration of policy and market processes that support decarburization. During this time, enabling technologies such as ESS, fuel cells, V2G frameworks, and smart grid infrastructure underwent significant evolution, increasing their usefulness in real-world deployments [28]. From a starting pool of 150 documents, 80 high-quality and thematically relevant sources were chosen for in-depth research. These sources were organized into main subject groups to aid in targeted examination and synthesis. The first theme focused on the integration of RESs with microgrids [29, 30], namely, energy management systems (EMS) [31], control techniques, and the effect of renewable intermittency on microgrid reliability [32]. Another focus was on EV charging infrastructure and how it interacts with the grid, including peak load management, charging station spatial distribution, and grid effect studies [33]. The study also looked at bidirectional charging technologies like V2G and V2H, examining how energy may be transferred between EVs and the grid to improve demand flexibility, lower peak loads, and improve overall system performance [34].

Fuel cells were assessed for their ability to supplement ESS and RES by offering a consistent, clean energy source that can operate regardless of weather conditions. Their capacity to generate power via electrochemical reactions, particularly hydrogen fuel cells, makes them a potential option for improving the stability and flexibility of microgrid systems, particularly in remote or high-demand areas [35]. Microgrid-enabled EV charging systems were assessed for their capacity to improve the charging infrastructure's sustainability, dependability, and operational efficiency.

These systems use localized renewable production, storage, and fuel cell technologies to reduce reliance on centralized grids and lower the environmental effect of vehicle electrification [36, 37]. A comparative methodology was used to investigate various system setups and integration strategies. This included investigating several RES-microgrid configurations, EV charging tactics, and technology synergies [38]. The study includes a comprehensive assessment of important performance characteristics such as cost-effectiveness, operational scalability, energy efficiency, and system reliability. This comparative review yielded best practices for effectively integrating renewable energy, microgrids, and EV infrastructure. To guarantee that the findings were legitimate and relevant, experts in EV technology, smart grid development, and renewable energy systems were consulted [39, 40]. Their feedback helped to refine the analytical framework and improve the evaluation of both technical performance and implementation tactics. These conversations also helped to verify and contextualize the conclusions in light of ongoing developments in the energy and transportation sectors. The analysis identified numerous areas where current research is insufficient or fragmented. Key limitations were highlighted in the development of scalable microgrid designs, improved control algorithms, and unified policy frameworks to facilitate the coordinated deployment of integrated systems [41]. Addressing these deficiencies requires future research to be systematic and methodical. This includes comparing the current research landscape, particularly in energy optimization, grid stability, and EV infrastructure performance with emerging demands such as system scalability, decarburization targets, and cross-sector interoperability [42].

Furthermore, unexplored areas like lifecycle cost analysis, interoperability between systems and technologies, and regional differences in deployment outcomes require special consideration. Existing research can be categorized by methodology such as simulations versus real-world trials and main focus areas including dependability, cost, and energy performance, which might reveal discrepancies or limits, such as a lack of full validation in operational situations. This study presents a set of focused recommendations to direct future studies after finding and articulating these research gaps. These proposals aim to spur innovation and promote the creation of robust, efficient, and sustainable energy ecosystems that seamlessly incorporate renewable generation, microgrid infrastructure, fuel cells, and EV technologies. Such developments are critical for achieving long-term energy and climate goals while maintaining practical viability and system-level reliability in a variety of geographic and regulatory contexts [43, 44].

3 Renewable Energy Sources

RESs are crucial in the transition to more sustainable and environmentally friendly energy systems. In contrast to conventional fossil fuels, which are finite and contribute significantly to greenhouse gas emissions and ecological degradation, RESs such as sun, wind, hydropower, biomass, and developing technologies such as fuel cells

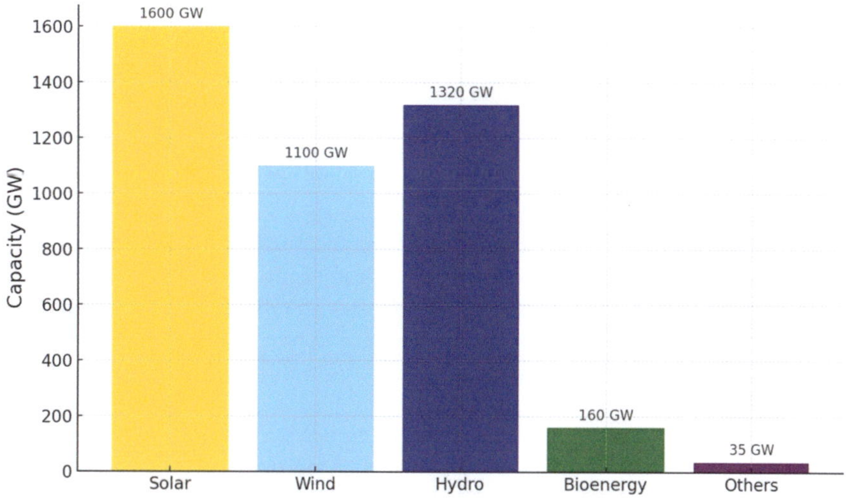

Fig. 1 Total renewable capacity of India by 2025

provide clean, abundant, and continually replenished sources of energy. Their incorporation into modern power networks is critical for lowering carbon footprints and ensuring long-term energy security. RESs are crucial in the transition to more sustainable and ecologically responsible energy systems. Unlike conventional fossil fuels, which are finite and large contributors to greenhouse gas emissions and environmental degradation, RESs such as sun, wind, hydropower, biomass, and developing technologies such as fuel cells provide clean, abundant, and continually renewed energy. Their incorporation into current power grids is critical for lowering carbon footprints and ensuring long-term energy security [45]. Renewable energy capacity is predicted to continue to rise rapidly in 2025 as shown in Fig. 1. Further, the different types of RESs are explained below.

3.1) **Wind energy** is another major renewable source that is produced by transforming the kinetic energy of wind into electricity using wind turbines. Wind farms can be built onshore or offshore, with offshore sites often experiencing higher and more regular wind speeds. While wind energy is clean and scalable, its viability is affected by geography, and there are worries about visual impact and animal implications [46].

3.2) **Solar energy** is one of the most well-known examples, as it harnesses sunlight either through PV cells that convert light directly into electricity or through concentrated solar power systems that generate thermal energy using mirrors or lenses. Solar thermal collectors are also utilized for water and room heating. Solar electricity is clean and abundant, but its intermittent nature necessitates storage or backup systems for reliability [47].

3.3) **Hydropower**, also known as hydroelectric energy, uses the flow of water—usually from rivers or dams to spin turbines and generate electricity. It is a well-established and dependable form of renewable energy, especially in areas with plentiful water resources. Hydropower systems can range from large-scale dam projects to smaller run-of-river arrangements that generate power without requiring much water storage. Large installations, on the other hand, have the potential to destroy ecosystems and displace populations [48].

3.4) **Biomass energy** is the utilization of organic materials, such as wood, agricultural wastes, and animal waste, to generate heat, electricity, or biofuels. Biomass can be burned directly, converted to petrol (biogas), or refined into liquid fuels such as ethanol and biodiesel. While it has the potential to be carbon-neutral, sustainable sourcing and emissions management are critical for avoiding environmental damage and competition with food production [49].

3.5) **Geothermal energy** is derived from the earth's internal heat. This energy can be obtained by tapping into subsurface pools of steam or hot water and generating electricity or providing direct heating. Geothermal heat pumps are also used to adjust building temperatures by transporting heat from the ground to the inside. Geothermal systems are noted for their great reliability and low emissions, but their feasibility is dependent on geological conditions [50].

3.6) ***Challenges and Integration Strategies for RESs***: The integration of RESs into microgrids and EV charging infrastructure presents some technical, operational, and economic issues, the majority of which come from the intermittent and variable character of sources such as solar and wind. These oscillations have a profound impact on the stability and reliability of localized energy networks.

3.6.1) Ensuring grid stability is of utmost importance as shifting power generation has the potential to create voltage and frequency fluctuations within microgrids, thereby jeopardizing continuity of supply and operation of various devices. In order to address these anomalies, dependable energy storage solutions such as lithium-ion batteries, flow batteries, and super capacitors are essential. These devices capture surplus energy during periods of low demand and dispense it when energy demand is high and the supply low, thus creating constant energy supply.

3.6.2) Moreover, the combination of different RESs requires more sophisticated power electronics, as well as autonomous control systems to manage the energy flow, handle dynamic loads, and interface with the primary power grid as well as EV charging systems.

3.6.3) These technologies allow real-time monitoring, balancing of energy, and fault identification which is crucial for the enhancement of the system performance as a whole. Also, there is an economic and regulatory rationale that impacts the pace and level of investment in RESs.

3.6.4) There is an explanation for the imbalance in the market's supply and demand when it comes to renewable energy sources. The intricate hierarchical relations make situation more complicated because of the elevated expenditures of installing and maintaining renewable energy sources. The changes in the market structure calls for policies, financial arrangements, and legislation that focus on strengthening renewable in order to promote investment and innovation.

4 Role of Microgrids in Renewable Energy and EV Integration

Integrating RESs with EV charging infrastructure would be impossible without microgrids which optimize the generation and consumption of energy on a smaller scale. They provide an adaptable base for the incorporation of various renewable energies like wind and solar to local energy systems. Through the use of ESS, microgrids become able to supply power consistently and dependably by managing the intermittency and unpredictability of renewable energy coupled with advanced control techniques. When connected to an EV charging station, microgrids can provide clean, renewable electricity for vehicle charging. Smart charging solutions, such as demand response and time-of-use pricing, allow microgrids to optimize energy usage by balancing supply and demand while reducing grid strain.

By the end of 2025, it is expected that roughly 500,000 electric cars (EVs) will be on Indian roads. Assuming an average energy consumption of 0.2 kWh per kilometer and an annual travel distance of around 10,000 km per car, each EV is expected to consume approximately 2,000 kWh of electricity each year. When combined with the complete EV fleet, this amounts to a total power consumption of around 1,000 GWh (or 1 TWh) for 2025 as shown in Fig. 2. Furthermore, bidirectional V2G technology enables EVs to function as mobile energy storage, returning electricity to the microgrid during peak demand or low RES generation periods. This improves grid stability, encourages efficient energy use, and facilitates the smooth introduction of renewable energy into the transportation sector [49].

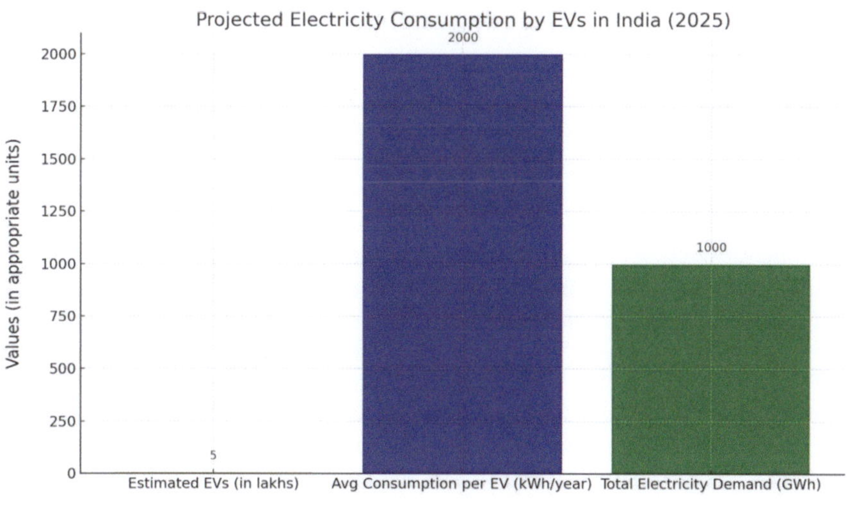

Fig. 2 Estimated EV electricity consumption by 2025 in India

5 EV Charging Infrastructure

The rapid rise of EVs is reworking the transportation sector, imparting a purifier and more sustainable opportunity to standard inner combustion engines. As EV utilization grows, the development of green and scalable EV charging infrastructure has end up a critical facilitator of big EV deployment. A well-designed charging community not only best improves patron convenience but also contributes substantially to grid balance and sustainability. EV charging infrastructure allows to make the electricity grid greater resilient and adaptable with the aid of integrating RESs, optimizing power use, and helping superior technology along with V2G structures. This phase seems at the numerous styles of EV chargers, the predominant components of charging infrastructure, the issues of scaling up EV charging networks, and how clever charging and microgrid integration can enhance the efficiency and efficacy of EV charging systems [51].

5.1) ***Types of EV Chargers and Charging Levels*** [52]

EV chargers are classed into three classes primarily based on charging speed and electricity output, with each degree designed for particular software.

5.1.1) Level 1 charger utilizes a standard 120 V AC outlet offering a maximum power of 2.4 kW which yields 3–5 miles of range per hour. They are predominantly used at homes and are perfect for low-range drivers who require overnight charging.

5.1.2) Level 2 chargers use a 240 V AC input furnishing power from 7.2 to 19.2 kW within 10–60 miles per hour dynamically sparing. These are usually installed in residences, workplaces, and other business institutions in addition to public garages and serve consumers who need routine quick charging.

5.1.3) Commonly referred to as Level 3 chargers, DC fast chargers apply direct current in the range of 50–350 kW. They are mostly found on highways, at service stations, and in busy city areas for ease of long-distance travel and high-demand charging.

5.2) ***Components of EV charging infrastructure*** [53]

5.2.1) Charging Stations

The most basic interaction that could be established involves charging stations. These structures contain the tools and terminals needed to provide energy for vehicles. There are various types of proprietary and commercial, from single outlet home units to larger multi-outlet units. Public charging stations often include payment and access control systems for safety and ease of use.

5.2.2) Electrical Vehicle Supply Equipment (EVSE)

The EVSE serves as the bridge between the power supply and the EV features, managing the electric power flow while ensuring system security and communication. It contains essential components such as communication circuits and interfacing relays that break with the onboard system and the power grid, which integrate with the vehicle system flexibly.

5.2.3) Power Conversion Equipment

The electricity meant for EV batteries is poured through inverters and rectifiers. These devices change the AC power from the grid to DC electricity, which most EVs use, especially at Level 2 and in DC fast charging systems, ensuring compatibility and charging efficiency.

5.2.4) Energy Storage System

Batteries in the form of ESS units are added to the charging stations and enable the storage of electricity from the grid or from RESs. They help control energy demand by supplying stored power at peak times and acting as a backup during outages. This contributes to grid stability and resilience. Permanently installed energy storage systems enhance reliability.

5.2.5) Renewable Energy and Fuel Cell Integration

Along with the EV charging infrastructure, PV systems and hydrogen fuel cells can be integrated as RESs. Fuel cells generate power by a chemical reaction between hydrogen and oxygen while emitting only water vapor. Solar PVs generate clean electricity. The use of these methods enhances the sustainability of EV charging by lessening the dependency on fossil fuels and decreasing greenhouse gas emissions.

5.2.6) Communication and Control Systems

Today's charging stations come equipped with highly sophisticated communication and control mechanisms that facilitate instantaneous communication between the automobile, the charger, and the power grid infrastructure. These offer smart grid features such as control of energy use, demand response management, and pricing schemes for energy use, thus optimizing energy distribution and enhancing grid performance.

6 Key Challenges in the Development of EV and Fuel Cell Charging Infrastructure

This section describes the key challenges arises during the establishment of fuel cell-based hybrid EV charging infrastructure.

6.1) ***Managing Grid Load and Ensure Stability***

The rapid rise in electric car use presents considerable difficulties to power system reliability. High-volume, simultaneous charging, especially during peak hours, can put a strain on local grids, causing voltage fluctuations, congestion, and even outages. To avoid these hazards, smart grid technology, demand-side management, and real-time communication between charging systems and utility operators must all be integrated [54].

6.2) ***Addressing High Installation and Operational Costs***

Both EV charging stations and hydrogen refueling facilities demand substantial capital expenditure. DC rapid chargers necessitate costly electrical improvements and grid connections, whereas hydrogen stations require specialized equipment for storage, compression, and dispensing, as well as safety precautions. Economic feasibility is dependent on financial incentives, legislative backing, and scalable business models that promote infrastructure deployment for both technologies [55].

6.3) ***Promoting Standardization and Interoperability***

A key impediment to a seamless user experience is the lack of standardization in connectors, communication protocols, and fueling procedures. Inconsistent charging standards for EVs can make them incompatible with different vehicle models and networks. Similarly, fuel cell infrastructure must be compatible with developing hydrogen purity and refueling regulations. Establishing standardized technical specifications is crucial for maintaining interoperability, safety, and convenience of use across all vehicle categories [56].

6.4) ***Optimizing Site Selection and Improving Accessibility***

Strategic placement of EV charging stations and hydrogen refueling stations is critical for increasing convenience and uptake. Land availability, zoning regulations, grid or pipeline access, and customer demand predictions are all potential challenges. Fuel cell plants, in particular, necessitate additional safety permits and specialized site planning. A comprehensive approach involving urban planning, stakeholder engagement, and geographic equity is necessary to expand accessible infrastructure for both EVs and FCEVs [57].

6.5) ***Smart Charging Technologies and Their Role in Modern EV Infrastructure***

Smart charging, also known as intelligent or managed charging, is a critical innovation in the development of electric transportation infrastructure. It uses advanced communication, control systems, and data analytics to optimize the charging process depending on real-time grid circumstances, dynamic electricity prices, energy availability, and user preferences. Smart charging promotes grid stability, cost efficiency, and environmental sustainability by aligning charging behavior with overall energy system requirements [51].

6.6) ***Dynamic Load Management and Demand Response***

All technologies V2G and V2H are fundamental for both EV and FCEV infrastructure. During peak hours, vehicles can use V2G to discharge electricity back to the grid, helping to balance the system. With V2H technology, the vehicle can also power home appliances during an outage or when electricity prices are high. Such an energy flow can establish energy resilience for FCEVs by enabling them to operate as energy resources and storage units that can be easily integrated with renewable energy systems [22].

6.7) ***V2G and V2H Technologies***

V2G and V2H technologies are critical to both EV and FCEV infrastructures. Vehicles can use V2G to provide electricity back to the grid during moments of heavy demand, which helps to stabilize the system. Similarly, V2H technology enables automobiles to power household appliances during power outages or spikes in electricity rates. This bidirectional energy flow can help FCEVs achieve energy resilience by allowing them to function as both energy sources and storage units, effortlessly integrating with renewable energy systems [35].

6.8) ***Integration with Microgrids and DERs***

The optimization of energy efficiency is achieved by integrating EV and FCEV charging architecture with microgrids powered locally by RESs such as solar, wind, or hydrogen produced by electrolysis. These microgrids can act as backup sources of power for the EV and FCEV charging stations during outages and, in doing so, decrease dependence on the primary power grid. This integration is especially beneficial for fuel cell stations, which frequently rely on hydrogen production that may be nearby supplied and stored, resulting in a resilient and efficient energy system [58, 59].

6.9) ***Role of EMS and Internet of Things (IoT) in Smart Charging***

The management of EV and FCEV charging is best accomplished using EMS and IoT technologies. These systems facilitate the dynamic adjustment of charging deadlines based on grid opportunities such as electricity cost, energy availability, and overall grid conditions. Moreover, they aid in stream scheduling for multiple sites which include both EV and FCEV chargers, improving energy distribution and RESs integration with FCEV and EV charging stations [55, 57].

6.10) ***Cybersecurity and Data Privacy in Smart Charging***

Robust cybersecurity and data confidentiality measures are essential as the smart charging network becomes more interconnected. Protecting user data and energy network integrity becomes a critical task for the advanced communication systems associated with EV and FCEV charging systems. Dynamic and modern changing energy markets demand safeguarding personal data from cyberthreats using secure communication methods, encryption, and compliance with privacy law. Along with V2G and V2H and microgrid integration, smart charging is a pioneering strategy to enhance charging—for both fuel cell and electric vehicles. These technologies not only optimize the user experience and deliver economic advantages to vehicle owners, but also foster a cleaner, more flexible, and resilient energy system [60, 61].

6.11) ***Future Trends and Opportunities***

The future of automotive charging infrastructure will evolve with the smart charging stations, their integration with RESs, and advancements like V2G systems, wireless charging, and microgrids. The ultra-fast DC chargers in cities and along the highways will help in long-distance traveling by significantly reducing charging durations. As components of smart city grid systems, EV and FCV charging infrastructure will further synergize with regard to intelligent transport systems, IoT systems, and energy management systems for efficient servicing and energy consumption optimization. The speed and ease of use provided by automated and self-driven charging solutions will further maximize convenience. Renewable sources like solar, wind, and green hydrogen, along with microgrids for local generation and resilience, will provide clean energy to battery electric and fuel cell vehicles. EVs and FCVs will serve as mobile energy assets with V2G and next-generation energy management features supporting flexible grid demand, bidirectional energy flow, increasing dispatchable power resources, flexibility, and load balancing. All these innovations will reduce emissions and energy system disproportionality; promote sustainable mobility; and contribute toward a cleaner, smarter, and more resilient energy framework.

7 Conclusion

The integration of microgrids with RESs, along with the charging facilities for EVs and FCEVs, marks a significant shift toward achieving a truly resilient, low-carbon, and sustainable energy system. This chapter has elaborated on the synergistic advantages of this integration, including increased renewable energy consumption, reduction in greenhouse gases and emission, improved reliability of the grid, and greater energy efficiency. Smart charging and advanced energy management systems integrated into microgrids enable precise control over the variability presented by RESs. Furthermore, bidirectional technologies such as V2G enhance the capability of EVs to provide auxiliary grid services and energy flexibility. Solid-state batteries, next-generation lithium-ion batteries, and other second-use EV battery systems for stationary storage enhance grid resiliency and renewable resource integration. Also, enhanced fuel cell technology combined with green hydrogen production offers an optional pathway for zero-emission vehicles, especially for long-range and heavy-duty applications where FCEVs offer faster refueling and longer travel distances. However, achieving the full potential of this integrated system necessitates addressing critical technical, economic, and regulatory issues. These include controlling the intermittent output of RESs, ensuring system interoperability, developing cost-effective storage and hydrogen infrastructure, and creating supporting governmental frameworks. Overcoming these issues will necessitate ongoing innovation, investment, and collaboration among governments, industry leaders, and research organizations. As energy systems grow, the convergence of renewable energy, microgrids, EV/FCEV infrastructure, and advanced control technologies will be critical to decarburizing both the energy and transportation sectors. This integration, with the proper strategic direction and legislative backing, has the potential to accelerate the transition to a cleaner, more efficient and resilient energy future.

References

1. Kumar N, Saxena V, Singh B, Panigrahi BK (2023) Power quality improved grid-interfaced PV-assisted onboard EV charging infrastructure for smart households consumers. IEEE Trans Consum Electron 69(4):1091–1100. https://doi.org/10.1109/TCE.2023.3296480
2. Kundu S, Singh M, Giri AK, Kadiyan S, Chittora P, Shailly D (2024) Robust and fast control approach for islanded microgrid system and EV charging station applications. Electr Eng. https://doi.org/10.1007/s00202-024-02291-1
3. Shukl P, Singh B (2023) Distributed energy resources based EV charging station with seamless connection to grid. IEEE Trans Ind Appl 59(3):3826–3836. https://doi.org/10.1109/TIA.2023.3239583
4. Kundu S, Giri AK, Kadiyan S, Singh S (2025) Power management and adaptive control method for a standalone distributed generation system under fault, DC-offset, and dynamic environments generation system under fault, DC-offset, and dynamic environments. https://doi.org/10.1080/01430750.2025.2452245

5. Chakraborty S, Modi G, Singh B, Panigrahi BK (2023) Control of a grid-tied battery-less two-stage solar PV system with seamless transition and autonomous dispatch capability. IEEE Trans Ind Appl 59(6):7669–7680. https://doi.org/10.1109/TIA.2023.3300289
6. Sorgulu F, Dincer I (2018) A renewable source based hydrogen energy system for residential applications. Int J Hydrogen Energy 43(11):5842–5851. https://doi.org/10.1016/j.ijhydene.2017.10.101
7. Giri AKK, Arya SR, Maurya R, Babu BC (2018) Power quality improvement in stand-alone SEIG-based distributed generation system using lorentzian norm adaptive filter. IEEE Trans Ind Appl 54(5):5256–5266. https://doi.org/10.1109/TIA.2018.2812867
8. Kundu S, Giri AK, Kadiyan S (2024) An adaptive mixed-step size normalized least means fourth control approach for standalone power generation system considering dynamic conditions. IEEE J Emerg Sel Top Power Electron 1. https://doi.org/10.1109/JESTPE.2024.3516878
9. Chishti F, Murshid S, Singh B (2020) Robust normalized mixed-norm adaptive control scheme for PQ improvement at PCC of a remotely located wind-solar PV-BES microgrid. IEEE Trans Ind Inf 16(3):1708–1721. https://doi.org/10.1109/TII.2019.2923641
10. Erdogan N, Pamucar D, Kucuksari S, Deveci M (2023) A hybrid power heronian function-based multicriteria decision-making model for workplace charging scheduling algorithms. IEEE Trans Transp Electrif 9(1):1564–1578. https://doi.org/10.1109/TTE.2022.3186659
11. Sharma R, Singh B, Kewat S (2022) Improved generalised integrator for optimal dispatch of SyRG-based pico-hydro-PV system. IET Energy Syst Integr 4(1):116–130. https://doi.org/10.1049/esi2.12052
12. Jaman S, Abdel-Monem M, Geury T, Hegazy O (2023) Development and validation of an integrated EV charging station with grid interfacing inverter for residential application. IEEE Access 11(September):115751–115774. https://doi.org/10.1109/ACCESS.2023.3323219
13. Kundu S, Singh M, Giri AK (2023) SPV-wind-BES-based islanded electrical supply system for remote applications with power quality enhancement. Electr Eng. https://doi.org/10.1007/s00202-023-01979-0
14. Meena G, Singh VP, Sundariyal N, Mathur A (2024) Component model of AC microgrid for short-circuit analysis. In: Proceedings of 2024 1st international conference on innovative sustainable technologies for energy, mechatronics, and smart systems ISTEMS 2024, pp 1–6. https://doi.org/10.1109/ISTEMS60181.2024.10560198
15. Sombir S, Singh M (2020) Voltage and frequency control of self excited induction generator integrated with PV system. In: IECON Proceedings (industrial electron configuration, pp 4306–4311. https://doi.org/10.1109/IECON43393.2020.9254608
16. Kundu S, Singh M, Giri AK (2022) Control algorithm for coordinated operation of wind-solar microgrid standalone generation system. Energy Sources Part A Recover Util Environ Eff 44(4):10024–10044. https://doi.org/10.1080/15567036.2022.2143943
17. Singh N, Pratap B, Swarup A (2019) Design of dynamic surface controller for robust performance of variable speed wind turbine. IET Renew Power Gener 13(9):1511–1521. https://doi.org/10.1049/iet-rpg.2018.5922
18. Arena G, Chub A, Lukianov M, Strzelecki R, Vinnikov D, De Carne G (2024) A comprehensive review on DC fast charging stations for electric vehicles: standards, power conversion technologies, architectures, energy management, and cybersecurity. IEEE Open J Power Electron 5(October):1573–1611. https://doi.org/10.1109/OJPEL.2024.3466936
19. Wu C, Zhou D, Ma Y, Sui Q, Lin X, Wei F (2023) Restoration enhancement for distribution systems based on multiple battery swapping stations. IEEE Trans Ind Appl 60(2):2025–2036. https://doi.org/10.1109/TIA.2023.3327992
20. Singh N, Pratap B, Swarup A (2018) Robust control design of variable speed wind turbine using quantitative feedback theory. Proc Inst Mech Eng Part A J Power Energy 232(6):691–705. https://doi.org/10.1177/0957650917751463
21. IEEE (2003) IEEE 1547-2003 standards. 1547 TM IEEE standard for interconnecting distributed resources with electric power systems standard
22. Güven AF, Ateş N, Alotaibi S, Alzahrani T, Amsal AM, Elsayed SK (2025) Sustainable hybrid systems for electric vehicle charging infrastructures in regional applications. Sci Rep 15(1):4199. https://doi.org/10.1038/s41598-025-87985-7

23. Kundu S, Singh M, Giri AK (2022) Implementation of variable gain controller based improved phase locked loop approach to enhance power quality in autonomous microgrid. Int J Numer Model Electron Netw Devices Fields, pp 1–19. https://doi.org/10.1002/jnm.3082
24. Yoon DH, Seo H, Lee J, Kim Y (2024) Online electric vehicle charging strategy in residential areas with limited power supply. IEEE Trans Smart Grid 15(3):3141–3151. https://doi.org/10.1109/TSG.2023.3339788
25. Kundu S, Giri AK, Singh M, Kadiyan S (2024) Adaptive control strategy for isolated renewable energy-based generation system with intermittency. Electr Eng. https://doi.org/10.1007/s00202-024-02454-0
26. Singh N, Pratap B, Swarup A (2019) Nonlinear robust observer based adaptive control design for variable speed wind turbine. J Eng Res 7(3):258–285
27. Samal KB, Pati S, Sharma R, Nayak AM (2022) Power control in an PV-fuel cell based hybrid isolated microgrid. In: Proceedings of-2022 IEEE 2nd international symposium on sustainable energy, signal processing and cyber security iSSSC 2022, pp 1–5. https://doi.org/10.1109/iSSSC56467.2022.10051266
28. Sombir, Singh M, Giri AK (2022) Implementation of in-loop filter based DSC-FLL control approach for power quality improvement in isolated microgrid. In: IEEE international symposium on industrial electronics, pp 516–521. https://doi.org/10.1109/ISIE51582.2022.9831712
29. Kumar S, Singh B (2019) Seamless operation and control of single-phase hybrid PV-BES-utility synchronized system. IEEE Trans Ind Appl 55(2):1072–1082. https://doi.org/10.1109/TIA.2018.2876640
30. Giri AK, Arya SR, Maurya R (2019) Compensation of power quality problems in wind-based renewable energy system for small consumer as isolated loads. IEEE Trans Ind Electron 66(11):9023–9031. https://doi.org/10.1109/TIE.2018.2873515
31. Verma A, Singh B (2020) Multimode operation of solar PV array, grid, battery and diesel generator set based EV charging station. IEEE Trans Ind Appl 56(5):5330–5339. https://doi.org/10.1109/TIA.2020.3001268
32. Kundu S, Singh M, Giri AK (2021) Design and control of a standalone wind-solar system with SEIG feeding linear/nonlinear loads. Lect Notes Electr Eng 616:317–327. https://doi.org/10.1007/978-981-16-1978-6_28
33. Mittal S, Singh A, Chittora P (2024) Power quality enhancement in single phase two level/five level converters using adaptive-RBFNN algorithm. Electr Eng. https://doi.org/10.1007/s00202-024-02432-6
34. Chauhan PJ, Chatterjee JK (2019) A novel speed adaptive stator current compensator for voltage and frequency control of standalone SEIG feeding three-phase four-wire system. IEEE Trans Sustain Energy 10(1):248–256. https://doi.org/10.1109/TSTE.2018.2832140
35. Savrun MM, İnci M, Büyük M (2021) Design and analysis of a high energy efficient multi-port dc-dc converter interface for fuel cell/battery electric vehicle-to-home (V2H) system. J Energy Storage 45(November):2022. https://doi.org/10.1016/j.est.2021.103755
36. Shalaby AA, Shaaban MF, Mokhtar M, Zeineldin HH, El-Saadany EF (2022) A Dynamic optimal battery swapping mechanism for electric vehicles using an LSTM-based rolling horizon approach. IEEE Trans Intell Transp Syst 23(9):15218–15232. https://doi.org/10.1109/TITS.2021.3138892
37. Rai KB, Kumar N, Singh A (2024) Design and control of DVR based on adaptive bateman polynomial for power quality improvement. Int J Circuit Theory Appl 1–16. https://doi.org/10.1002/cta.4270
38. Singh S, Chauhan P, Singh NJ (2021) Feasibility of grid-connected solar-wind hybrid system with electric vehicle charging station. J Mod Power Syst Clean Energy 9(2):295–306. https://doi.org/10.35833/MPCE.2019.000081
39. Systems C et al (2024) Integration of renewable energy , microgrids, and EV charging infrastructure : challenges and solutions, vol 2, no 3. https://doi.org/10.59247/csol.v2i3.142

40. Singh N, Kundu S, Verma V, Nangru Y (2025) Effective energy efficiency methods for distribution transformers. In: 1st international conference on advances in computer science, electrical, electronics, and communication technologies CE2CT 2025, pp 579–58. https://doi.org/10.1109/CE2CT64011.2025.10939316
41. Yan D, Yin H, Li T, Ma C (2021) A two-stage scheme for both power allocation and EV charging coordination in a grid-tied PV-battery charging station. IEEE Trans Ind Inf 17(10):6994–7003. https://doi.org/10.1109/TII.2021.3054417
42. Mittal S, Singh A, Chittora P (2024) Design and development of leaky least mean fourth control algorithm for single-phase grid-connected multilevel inverter. Int J Circuit Theory Appl 52(1):328–345. https://doi.org/10.1002/cta.3746
43. Chauhan S, Lingaiah ABN, Tummuru NR (2023) Power management of renewable-grid integrated smart e-mobility system for light electric vehicles. IEEE Trans Transp Electrif 9(3):4183–4195. https://doi.org/10.1109/TTE.2023.3238897
44. Rai KB, Kumar N, Singh A (2024) Three-phase grid connected shunt active power filter based on adaptive Q-LMF control technique. IEEE Trans Power Electron 39(8):10216–10225. https://doi.org/10.1109/TPEL.2024.3398369
45. Narayanan V, Singh B (2023) Seamless operation of an SRG based WEGS and a bidirectional converter interfaced BES microgrid under polluted grid conditions. IEEE Trans Ind Appl 1–12. https://doi.org/10.1109/TIA.2023.3306741
46. Qureshi A, Giri AK, Raj S, Padmanaban S (2022) Power conditioning using DSTATCOM in a single-phase SEIG-based isolated system. Electr Eng 104(1):111–127. https://doi.org/10.1007/s00202-021-01423-1
47. Pavan Kumar AV, Parimi AM, Uma Rao K (2015) Performance analysis of a two-diode model of PV cell for PV based generation in MATLAB. In: Proceedings of 2014 IEEE international conference on advanced communications, control and computing technologies, ICACCCT 2014, no 978, pp 68–72. https://doi.org/10.1109/ICACCCT.2014.7019191
48. Kewat S, Singh B (2020) Improved reweighted zero-attracting quaternion-valued LMS algorithm for islanded distributed generation system at im load. IEEE Trans Ind Electron 67(5):3705–3716. https://doi.org/10.1109/TIE.2019.2916301
49. Karmaker AK, Hossain MA, Pota HR, Onen A, Jung J (2023) Energy management system for hybrid renewable energy-based electric vehicle charging station. IEEE Access 11(March):27793–27805. https://doi.org/10.1109/ACCESS.2023.3259232
50. Hüner B (2024) Feasibility and environmental analysis of biogas-based hybrid energy system using HOMER pro software: a case study for Hatay. Energy Convers Manag 326:119480. https://doi.org/10.1016/j.enconman.2025.119480
51. Barman P et al (2023) Renewable energy integration with electric vehicle technology: a review of the existing smart charging approaches. Renew Sustain Energy Rev 183:113518. https://doi.org/10.1016/j.rser.2023.113518
52. Zhang D, Huber J, Kolar JW (2023) A three-phase synergetically controlled buck-boost current DC-link EV charger. IEEE Trans Power Electron 38(12):15184–15198. https://doi.org/10.1109/TPEL.2023.3300691
53. Jain V, Singh B, Seema (2023) A grid connected PV array and battery energy storage interfaced EV charging station. IEEE Trans Transp Electrif 9(3):3723–3730. https://doi.org/10.1109/TTE.2023.3234994
54. Wu H (2022) A survey of battery swapping stations for electric vehicles: operation modes and decision scenarios. IEEE Trans Intell Transp Syst 23(8):10163–10185. https://doi.org/10.1109/TITS.2021.3125861
55. Adegbohun F, von Jouanne A, Lee KY (2019) Autonomous battery swapping system and methodologies of electric vehicles. Energies 12(4):1–14. https://doi.org/10.3390/en12040667
56. Nimalsiri NI, Mediwaththe CP, Ratnam EL, Shaw M, Smith DB, Halgamuge SK (2020) A survey of algorithms for distributed charging control of electric vehicles in smart grid. IEEE Trans Intell Transp Syst 21(11):4497–4515. https://doi.org/10.1109/TITS.2019.2943620
57. Saxena N, Hussain I, Singh B, Vyas AL (2018) Implementation of a grid-integrated PV-battery system for residential and electrical vehicle applications. IEEE Trans Ind Electron 65(8):6592–6601. https://doi.org/10.1109/TIE.2017.2739712

58. Wang P et al (2024) A novel cooperative control for SMES/battery hybrid energy storage in PV grid-connected system. IEEE Trans Appl Supercond 34(8):1–5. https://doi.org/10.1109/TASC.2024.3420310
59. Kundu S, Singh M, Giri AK (2024) Synchronization and control of WECS-SPV-BSS-based distributed generation system using ICCF-PLL control approach. Electr Power Syst Res 226:109919. https://doi.org/10.1016/j.epsr.2023.109919
60. Alanazi F (2023) Electric vehicles: benefits challenges and potential solutions. J Appl Sci 13:1–23
61. Kundu S, Sundriyal N, Singh N, Kadiyan S, Singh S, Ramirez JM (2025) Control strategy for a single-phase grid-connected PV system with bidirectional EV charging support, modern computing technologies for ev efficiency and sustainable energy integration, pp 179–198. IGI Global Scientific Publishing. https://doi.org/10.4018/979-8-3373-2382-4.ch007

Chapter 2
Energy Storage Solutions and Induction Motor Drives for Sustainable EV Charging

Vaishali Rai, Arpit Yadav, Dillip Kumar Mishra, Manob Hazarika, Nikunj R. Patel, and D. P. Kothari

Abstract The rapid evolution of electric vehicles (EVs) needs advanced powertrain and charging technologies that emphasize efficiency, sustainability, and smart grid connectivity. This chapter provides an overview of the conjoining synergism of induction motor (IM) drive technology with power electronic-based charging infrastructures in the arena of next-generation electric vehicles (EVs). The discussion begins with the essential principles and advantages of using induction motors in the context of EV propulsion, with a primary focus on dynamic performances, favorable speed-sensorless control approaches, and regenerative braking. After an assessment of the power electronic interfaces, including inverter and motor controller technologies that bridge the IMs and EV power subsystem, we examine the dynamic evolution of the EV charging infrastructures, which includes both onboard and offboard charging systems, smart chargers, and the capabilities of offering bidirectional energy flow (e.g., vehicle-to-grid (V2G) and vehicle-to-home (V2H)). The core focus of the chapter emphasizes how these domains intersect and, together, enable the integrated connection between the drive and charge domains through shared power electronic interfaces, regenerative braking benefits, thermal heat management, and appropriate control strategies. Finally, the chapter covers industry standards, energy

V. Rai · N. R. Patel
Department of Electrical Engineering (EE-EVT), U.V.P.C.E., Ganpat University, Gujarat, India
e-mail: nrp05@ganpatuniversity.ac.in

A. Yadav (✉)
Department of Electrical Engineering, Madan Mohan Malaviya University of Technology, Gorakhpur, U.P, India
e-mail: ph20ee1003@cit.ac.in

D. K. Mishra
School of Electrical Engineering, KIIT University, Bhubaneswar, India

M. Hazarika
Department of Electrical Engineering, KIPM College of Engineering, Gorakhpur, U.P, India

D. P. Kothari
Indian Institute of Technology, Delhi, New Delhi, India

S. Kundu et al. (eds.), *Hybrid Renewable Power Infrastructure for Sustainable Electric Vehicle Development*, Studies in Infrastructure and Control,
https://doi.org/10.1007/978-981-95-1323-9_2

management strategies for integrated IM drives and charging infrastructures, and expected advancements in EV charging infrastructure and IM drive developments. Furthermore, the chapter also addresses the growing role of artificial intelligence and machine learning in optimizing system or network behaviors using EVs.

Keywords Induction motor drives · Power electronic converters · Electric vehicle charging infrastructure · AI and machine learning in EV control · Synergistic integration in EV systems

1 Introduction

The transition from the iconic internal-combustion vehicles to electric vehicles (EVs) represents one of the most notable technological and environmental changes of the twenty-first century. Although electric vehicles were developed more than 100 years ago, they frequently lay dormant for decades, being overshadowed by their gasoline-powered counterparts. With increasing international concerns about climate change, air quality in urban areas, and dependency on fossil fuels, EVs have regained traction as a critical line of defense in the decarbonization of transportation [1–3].

At the center of this electric revolution is the electric drive system—the motor and the control electronics—which transduces electrical energy to mechanical motion. In general, there are many varieties of motor technologies, but induction motors (IMs) have emerged as a competitive motor technology for EV propulsion. Initially, PMSMs outperformed IMs in terms of efficiency and power density, with PMSMs defining the state of the art in electric propulsion system performance, IMs have had a resurgence as a competitive motor technology driven by developments in power electronics, real-time control algorithms, and thermal management [3–5].

Induction motors have a number of useful advantages: they are robust, lower cost, do not require rare earth materials, and have a well-developed industrial design. Together with the advances in vector control, direct torque control, and sensorless control techniques, today's IM drives [12] can meet the performance of PMSMs in many EVs. IMs are inherently more robust under environmental conditions, making them appropriate for the many driving conditions experienced during everyday transport [5–7].

As propulsion system development has progressed, the supporting charging infrastructure for incentivizing EV adoption has also improved rapidly. Initially, slow grid interacting systems with limited intelligence were the only option for charging. Currently, the trend is docking systems that are fast, smart, open-ended, and bidirectional, and systems that interact with the grid, which are able to learn the user's behavior and the energy prices. Both onboard chargers and offboard chargers can now support power of many hundreds of kilowatts, and the newest versions embrace Vehicle-to-Grid (V2G) and Vehicle-to-Home (V2H), which allow EVs to not only receive power but to release power to the grid as needed. This capability allows EVs to be a more impactful resource rather than being a passive consumer of energy and allows them to become a valuable asset to the electricity networks as they maintain stability and offer services within demand response markets [7–9].

Motor drives and chargers share increasing functionality and hardware requirements, as manufacturers are now able to co-design these products to be synergistic. The most promising application to date is for power electronics in dual-use applications. In a specific application, the inverter used to power the induction motor can also be used for battery charging when the induction motor is not engaged. This collaboration not only decreases the BOM count and system complexity but also enhances efficiency and thermal management, particularly with shared heat sinks and cooling systems [10]. Additionally, regenerative braking allows the induction motor to act as a generator for deceleration, feeding energy back to the battery or grid. The regenerative capability relies heavily on the appropriate architecture of the charging system and should be co-designed for optimal energy flow, power quality, and system stability [11].

From a systems architecture standpoint, the integration of driving and charging electronics requires sophisticated energy management techniques to switch driving and charging roles. It requires intelligent controllers to make decisions on the fly and predict the future engagement of driving and charging modes. In this case, artificial intelligence (AI) and machine learning (ML) can be applied to predict energy demand, enable optimized routing from the perspective of charging, and, finally, predict component failures based on the prognostics of sensors placed within the drivetrain. There are challenges with integration, nevertheless. Thermal design becomes more important because there are more dual-purpose components. In addition, there is little standardization applied to the different EV platforms, which contributes to the difficulty in scaling and finding interoperability of shared infrastructure. These challenges need to be solved through collaborative standards, modular hardware, and adaptive software for all platforms [12].

To sum up, the potential for improved performance, extended range, and cost savings that are developed by converging both induction motor drives and charging infrastructure represents a substantial opportunity for the next generation of EVs. We plan to combine the strengths of the two, align the emerging induction motor drive (IMD) industry with aspects of the emerging charging infrastructure (CI) industry by developing a common hardware solution, regenerative control, and intelligent design and operation. This chapter will lay the groundwork for this emerging technology area by providing a formal review of this converged opportunity, connecting previous traditionally isolated systems, and initiating a path for more integrated designs to enable advanced sustainable transportation systems [13].

The principal aim of this chapter is to investigate and create an integration system that combines induction motor (IM) drives and power electronic-based charging infrastructure for future electric vehicles (EVs). In the context of the electric mobility market evolution, it highlights the significance of unifying hardware and controls between propulsion drives and charging systems to simplify designs, improve performance and reliability, and reduce costs. The chapter will identify potential efficiencies based on recent IM technology developments, vehicle-to-grid (V2G) controls and communications, intelligent control systems, and grounded on bidirectional converters to

- Present a holistic understanding of how IM drives and charging systems can be co-designed.
- Highlight the technical synergies, such as shared power electronics, regenerative braking, and intelligent energy management.
- Discuss standardization and scalability issues affecting interoperability.
- Outline the potential of AI and machine learning in enhancing operational coordination.
- Provide a strategic perspective for researchers, engineers, and policymakers to design unified, scalable EV architectures for the future (Fig. 1).

The chapter is divided into five sections, each of which addresses a particular element of the convergence. In Sect. 1, we cover an introduction, basics and benefits of induction motors for electric vehicles. Sections 2 and 3 describe both the power electronics and motor drive control and the charging infrastructure. Section 4 covers the hardware concept and intelligent control in future trends. Section 5 presents the conclusions.

Each section is designed to build toward an understanding of how converging propulsion and charging architectures will influence the future of electric mobility.

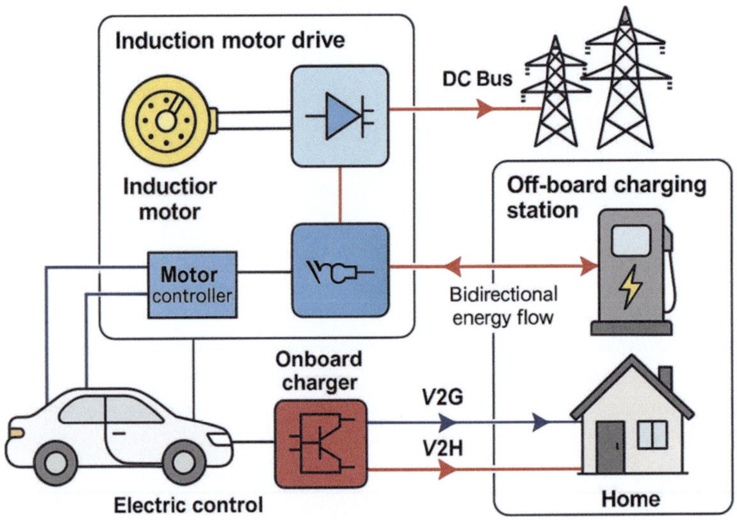

Fig. 1 Integrated architecture of induction motor drive and bidirectional charging system in electric vehicles [13, 14]

2 Power Electronics and Motor Drive Control

Power electronics form the technical foundation of vehicle propulsion electrification. This is especially true concerning induction motor (IM) drives. The fundamental components, inverters, rectifiers, and motor controllers are essential in power conversion and management. They enable the complex recycling and transformation of electrical energy by converting DC power from the battery to the variable frequency AC that an induction motor requires. In EV traction systems, inverters specifically provide an electrical interface that regulates the dynamic performance of the IM through several methods, such as vector control and pulse-width modulation. Additionally, inverters are designed so that torque production throughout the many operating conditions is efficient. Rectifiers can be located in onboard or offboard chargers, essentially performing the reverse conversion of AC grid power to regulated DC, which is required for storage in the onboard battery. Motor controllers are used as decision control units in which control algorithms regarding speed, torque, thermal limits, and regenerative braking are run while interfacing to the broader manage vehicle energy system [14, 15].

A significant trend over the last several years has been the establishment of dual-purpose converter architectures, a shared power electronics design that allows the same inverter used to power the induction motor to be used as a high-efficiency battery charger when the vehicle is parked. This has attractive benefits including reduced amount of hardware redundancy, reduced system costs and system weight. By removing the need for a dedicated charger circuit and maximizing component use, manufacturers could improve the performance of their vehicle powertrains while reducing production costs. But it does not come without challenges. For example, thermal management needs to be more involved since components are considered in dual modes, and this also identifies the need for improved techniques for heat dissipation mechanisms and more advanced control loops to avoid thermal issues, performance issues, or instability. Switching from a propulsion mode to a charging mode requires a real-time reconfiguration of the controller logic that must be speedy, fault-tolerant, and secure. Various factors and design considerations will warrant the signal integrity, isolation design and electromagnetic compatibility, and even more so in a high-voltage world [16–18].

In addition, since the vehicle connects to the grid and interacts with it through the two-way energy paths enabled by V2G and V2H, the power electronics system will also need to be able to interface with the grid. These interface requirements include synchronizing with the grid voltage and frequency, meeting grid codes and harmonic distortion thresholds, and including protection schemes. The motor drive and power electronics system is no longer just a closed-loop controller of motion; it is now this multi-role power interface, connecting propulsion, storage, and the grid infrastructure into one cohesive structure [20] (Fig. 2).

In conclusion, not only does the inclusion of power electronic converters and control systems improve the motor drive in EVs, but it also creates opportunities for multi-functionality, interoperability, and optimization at the system level. This

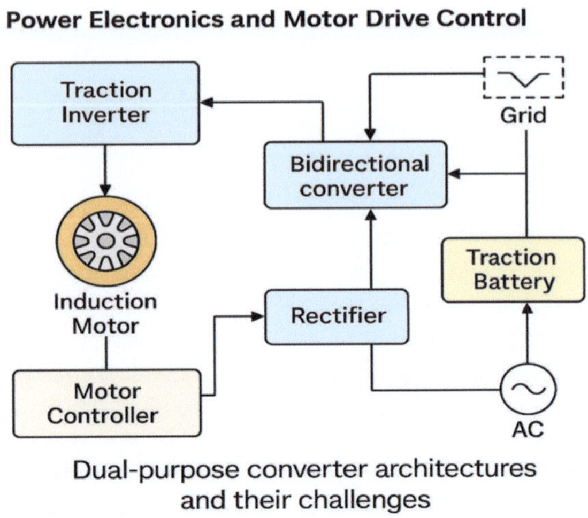

Fig. 2 Dual-purpose converter architectures and their challenges [16–19]

evolution is necessary for the larger vision described in the chapter, where propulsion and charging systems are designed in a synergistic EV architecture with intelligent resource efficiency [21].

3 Charging Infrastructure: Trends and Technologies

(onboard versus offboard charging, fast chargers, V2G/V2H technologies, grid impact, and standards like ISO 15118).

Electric vehicle (EV) charging infrastructure is continually evolving to accommodate the demands of rapid electrification and the requirements of smart energy management. Previously, charging infrastructure was limited to unidirectional AC charging stations with low charging power. Today, the EV charging environment has adopted multiple architectures, which include onboard and offboard charging systems. An onboard charging solution combines the AC-to-DC conversion hardware and software inside the vehicle, which is useful for home or workplace charging, given its flexibility and usability [22]. An onboard charging system is constrained by the vehicle's internal charger, so it is best for low-to-moderate charging power ratings (typically up to approximately 22 kW) and can manage the battery charging via an onboard vehicle charger. Offboard charging systems like DC fast chargers have the power conversion hardware located in the charging station, which enables power delivery from 50 to 350 kW (or more) and imposes minimal thermal constraints on the vehicle, allowing charging within minutes for long-distance and commercial requirements.

The next frontier in EV charging infrastructure is bidirectional energy flow, which is the premise for Vehicle-to-Grid (V2G) and Vehicle-to-Home (V2H) applications.

With V2G, electric vehicles can discharge electricity into the grid, acting as mobile energy storage assets, helping to balance supply–demand variations in real time by acting as active participants in smart grid operations instead of passive consumers, providing services such as peak shaving, frequency response, or renewable energy storage [23]. Similarly, V2H allows electric vehicles to act as a power source to households during grid outages or during points of peak energy pricing, bolstering energy independence and adding to grid resilience. These renewable energy innovations not only improve energy efficiency across the system but also create frameworks for new business models and incentives for electric vehicle owners.

Nevertheless, the rollout of smart and interoperable charging points presents major grid integration problems. With an increasing number of EVs, the amount of power consumed from the grid during times of peak demand may cause instability, voltage dips, or transformer overload unless managed effectively. Advanced load forecasting, dynamic tariffs, and smart charging algorithms are being adopted to solve these concerns. These technologies are intended to stagger, modulate, or delay charging sessions depending on grid conditions, user behavior, and renewable energy availability [24, 25].

Critically, the success of this interdependent ecosystem relies on sound communication standards, most notably ISO 15118. The global standard regulates the information exchange between EVs and charging points, allowing capabilities like Plug & Charge (automatic billing and authentication), encrypted messaging, and data exchange for V2G operations. ISO 15118 allows for the seamless integration of EVs into a smart grid under a high-level communication protocol under the state of cybersecurity and interoperability across various manufacturers and platforms. It is central to achieving the vision of a unified, scalable, and intelligent EV charging ecosystem [26].

Overall, contemporary charging infrastructure has gone far beyond mere power transfer; it has evolved to become a highly developed, multi-faceted system that combines vehicle propulsion, battery management, user experience, and grid stabilization. Through the knowledge of onboard and offboard charger trade-offs, utilization of high-speed and bidirectional charging, and following global standards such as ISO 15118, stakeholders are able to develop and roll out charging systems that are not only scalable and efficient, but also complementary with motor drives in vehicles and emerging energy systems [27] (Fig. 3).

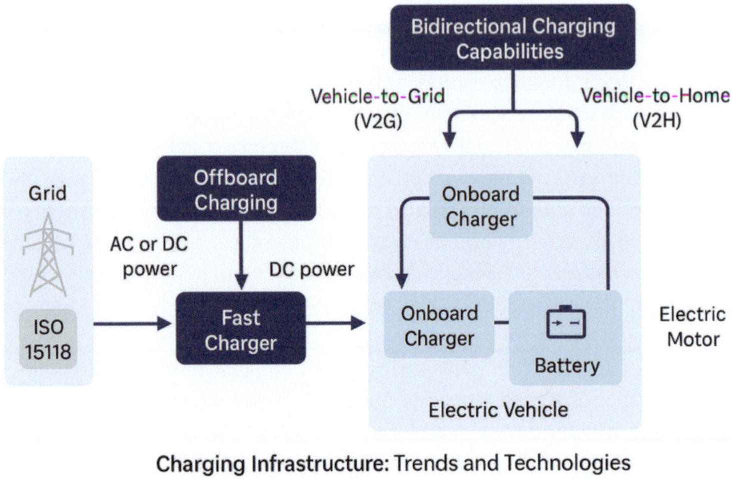

Fig. 3 Charging infrastructure: trends and technologies [27]

4 Integrated Control, Optimization, and Future Outlook for EV Drive-Charging Systems

4.1 Synergistic Integration of Motor Drives and Charging

The genuine promise of electric vehicles (EVs) is not merely in individual propulsion or charging subsystems but in synergistic unification of both. The new paradigm of dual-use architecture takes advantage of common hardware elements, including power electronic converters, in dual-mode operation—operating as both traction inverters while driving and battery chargers while stationary or plugged in. Such a combination eliminates redundant circuitry and maximizes thermal and spatial resources, leading to lighter vehicle weight, price, and complexity. The common hardware has to be dynamically reconfigured by means of intelligent control algorithms to meet varying voltage, current, and power demands during drive and charge cycles [28–30].

Another essential element of this synergy is regenerative braking combined with charging capability. Induction motors, running in generator mode while the vehicle slows down, are able to retrieve kinetic energy and return it to the battery or even the grid in V2G systems. This necessitates close coordination among the drive controller, onboard battery management, and energy storage subsystems to achieve high recovery efficiency while ensuring system safety and power quality. Further, such dual-use systems are optimally complemented by adaptive energy optimization techniques that proactively manage power flow in propulsion, braking, and charging modes, thus maximizing energy efficiency and efficient grid interaction [31].

4.2 Intelligent Control and Future Trends

With the increasing complexity of EV architectures, control mechanisms are taking center stage in ensuring seamless coordination of motor drives and charging infrastructure. The use of AI and ML algorithms enables predictive diagnostics, allowing for real-time identification of anomalies, component aging, and energy inefficiencies. These smart systems can dynamically adjust to a broad range of driving conditions, user preferences, and grid signals, learning the optimal strategies of motor control, energy distribution, and charge scheduling [32, 33].

In addition, the future of car control is headed toward cyber-physical systems (CPS), a tightly integrated universe of sensors, software, and actuators that are in concert with outside networks such as the power grid and the cloud. CPS enables real-time vehicle-state monitoring, energy demand forecasting, and predictive maintenance with real-time telemetry. Complementing this is the idea of digital twins, in which a computer simulation of the car simulates the behavior and performance of the physical system. This computer avatar facilitates simulation-driven optimization, where engineers can experiment with energy management, fault tolerance, and control efficiency algorithms before implementation [34].

The transition to software-defined EVs (SDEVs) wherein software controls not just infotainment and user interfaces but also powertrain dynamics, grid connectivity, and battery health is a revolutionary trend. As enabled by regular over-the-air (OTA) updates, manufacturers can tune for improved performance, deploy regulatory fixes, and incorporate new control strategies years after the car is in service [35–37]. This synergy of AI, CPS, and digital twinning underlies the development of adaptive, secure, and future-proof EV systems.

4.3 Challenges, Standardization, and Policy Implications

While the promising developments are encouraging, the journey to complete synergy between motor drives and charging systems is by no means without hurdles. Perhaps the most immediate concern is thermal design—because common components work under differential loads for different duties (driving, charging, regenerating), they need heavy-duty cooling and insulation schemes. Poor thermal management can result in component failure, lowered system efficiency, and safety risks [37, 38].

Interoperability is also another obstacle, especially in a disintegrated market with multiple platforms, power levels, and communication protocols. Without standardized protocols, scaling dual-use systems between manufacturers or integrating them with grid infrastructure is challenging. International initiatives like ISO 15118 are filling these gaps, but all-around consensus and implementation are needed to achieve complete compatibility [39, 40].

From a governance viewpoint, policy paradigms need to change to enable collective power electronics, grid services via V2G, and security of digital infrastructure.

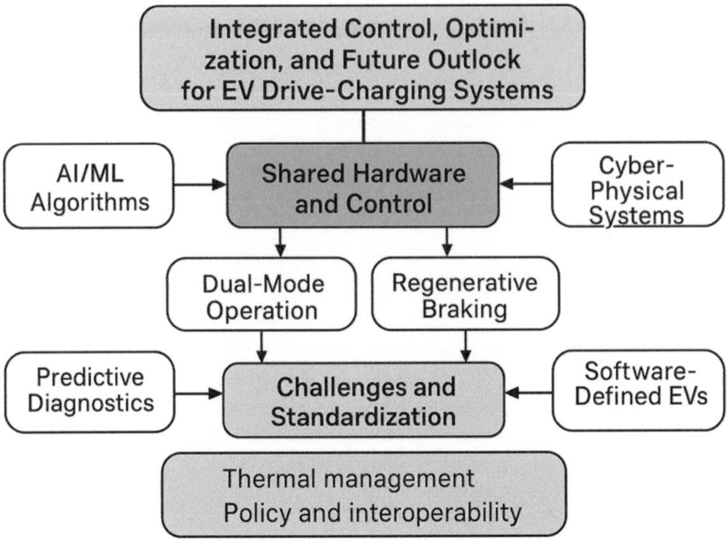

Fig. 4 Integrated control, optimization, and future outlook for EV drive-charging systems [37–39]

Incentives for implementing smart, synergistic architectures, such as tax incentives, R&D grants, or grid service compensation, can drive the implementation of these next-generation EV systems more quickly. Ultimately, ongoing research is needed in domains such as wide-bandgap semiconductors, fault-tolerant converter design, cybersecurity of CPS, and scalable ML algorithms for real-time energy optimization [41–43].

At its core, the journey to an integrated, intelligent EV ecosystem demands multidisciplinary co-operation, visionary regulations, and technologies bridging materials science, control engineering, AI, and power systems. Only in this concerted effort can the ultimate promise of synergistic integration be fulfilled for electric mobility (Fig. 4).

5 Conclusions and Future Scope

The synergistic unification of induction motor drives with power electronic-based charging infrastructure represents a key step in next-generation electric vehicle development. Through shared power electronic hardware, e.g., dual-mode inverters and integration of regenerative braking and bidirectional energy flow mechanisms (e.g., V2G/V2H), the integrated approach drastically increases efficiency, minimizes hardware redundancy, and decreases overall system cost. Incorporating AI- and ML-based control systems further allows real-time diagnostics, adaptive control, and smart energy optimization, coordinating EV operation with dynamic grid loads and user behavior.

Yet, this promising convergence comes with its set of challenges. Thermal management of dual-function components, the absence of universal interoperability standards, and regulatory gaps pose critical barriers to large-scale deployment. These need to be overcome through interdisciplinary collaboration within power electronics, communication protocols, control systems, and policy areas. ISO 15118 type standards need to be broadened and implemented extensively to facilitate seamless integration across varied EV platforms and infrastructures.

Going forward, the promise of electric mobility will be realized by deepening this integration through technology such as wide-bandgap semiconductors, software-defined EV structures, cyber-physical system modeling, and digital twin technologies. These technologies will enable greater levels of system efficiency, reliability, and intelligence. Moreover, the advent of decentralized energy systems and smart grid interactions makes EVs not just transport solutions but also mobile energy assets in an extended clean energy ecosystem.

Overall, the synergistic architecture of induction motor drives and charging infrastructure provides not only a technical upgrade but a strategic template for constructing smart, scalable, and sustainable transportation networks. Future work can aim to streamline this integration under practical conditions, encourage standardization, and open up new applications in energy management and grid services, keeping EVs ahead of the global energy curve.

References

1. Abd El-Raouf A, Elkholy MM, Elhameed MA, El-Arini M (2017) Effect of antlion optimized facts to enhance three-phase induction motor dynamic performance. In: 2017 Nineteenth international middle east power systems conference (MEPCON), pp 277–286
2. Aswal P, Gupta S, Joshi R, Gurung S, Surya KK (2016) Direct-axis calculation of direct torque control of induction motor using SVM. Int J Res Soc Sci 6(7):609–625
3. Baby LM, Salitha K (2013) Speed control of maximum boost controlled Z source converter fed induction motor drive with peak DC link voltage control. In: 2013 International conference on control communication and computing (ICCC), pp 281–286
4. Banerjee T, Choudhuri S, Bera J, Maity A (2010) Off-line optimization of PI and PID controller for a vector controlled induction motor drive using PSO. In: International conference on electrical and computer engineering (IEEE ICECE 2010), pp 74–77
5. Chen M, Zhang W (2015) H 2 optimal speed regulator for vector controlled induction motor drives. In: The 27th Chinese control and decision conference (2015 CCDC), pp 1233–1236. IEEE
6. Ak A, Rahmatullah R, Serteller NFO (2023) Design of sliding mode control using SVPWM modulation method for speed control of induction motor. Transp Res Proced 70:226–233
7. Laribi M, Ait Cheikh MS, Larbes C, Essounbouli N, Hamzaoui A (2013) A sliding mode and synergetic control approaches applied to induction motor. In: 3rd International conference on systems and control, pp 212–217. IEEE
8. Zaidi E, Djedoui L, Selmoune B, Mabrek A, Mezaache SE, Marouani K (2024) Nonlinear synergetic control of circulating currents in parallel three-phase inverters for power induction

motor drive. In: 2024 1st International conference on electrical, computer, telecommunication and energy technologies (ECTE-Tech), pp 1–6. IEEE
9. Zaidi E, Reffas A, Talhaoui H, Aissa O, Zouache R, Marouani K, Becherif M (2024) New PI-synergetic hybrid control for suppressing circulating currents of an electrical drive system fed by two parallel inverters. Electr Eng 1–19
10. Benbouhenni H, Bizon N (2021) A synergetic sliding mode controller applied to direct field-oriented control of induction generator-based variable speed dual-rotor wind turbines. Energies 14(15):4437
11. Li M, Chen Z, Mu H, Lun S (2023) Synergistic optimization design method to improve the overload capacity and efficiency of canned permanent magnet synchronous motor for vacuum pump. Electronics 12(22):4631
12. Nagesh Kumar GV, Veda Sai R, Mahitha Reddy T, Shajid SK, Boora K (2024) Emulation of wind turbine system using fuzzy controller-based vector controlled induction motor drive. J Mod Technol 38–46
13. Patil VJ, Khadake SB, Tamboli DA, Mallad HM, Takpere SM, Sawant VA (2024) Review of AI in power electronics and drive systems. In: 2024 3rd International conference on power electronics and IoT applications in renewable energy and its control (PARC), pp 94–99. IEEE
14. Zhang Y, Zhang Z, Babayomi O, Li Z (2023) Weighting factor design techniques for predictive control of power electronics and motor drives. Symmetry 15(6):1219
15. Lalitha SD, Danamaraju C, Prabhu Raj M, Kumar D, Sethuraman R (2024) AI-driven control strategies for power electronics converters. In: 2024 Ninth international conference on science technology engineering and mathematics (ICONSTEM), pp 1–6. IEEE
16. Saleh SA (2024) Development and performance testing of a control for permanent magnet synchronous motor drives with wavelet modulated power electronic converters. IEEE Trans Ind Appl 60(3):5025–5037
17. Kumar C, Kumar P, Raj N (2024) Performance comparison of PI and fuzzy logic controllers for speed control of permanent magnet sensorless brushless DC motors. In: 2024 IEEE 1st international conference on green industrial electronics and sustainable technologies (GIEST), pp 1–6. IEEE
18. Shahid MB, Jin W, Abbasi MA, Husain ARB, Munir HM, Hassan M, Flah A, Souissi ASE, Alghamdi TAH (2024) Model predictive control for energy efficient AC motor drives: an overview. IET Electric Power Appl 18(12):1894–1920
19. Davidson K, Kogalur N, Tolbert I, Watt E, Meintz A (2024) EVs@ scale-next-gen profiles: high level analysis and procedures Report. No. NREL/TP-5400–88898. National Renewable Energy Laboratory (NREL), Golden, CO (United States).
20. Thurston S, Wells L (2023) EVs@ scale next-gen profiles-EV profile capture 2023. No. ANL/TAPS-24/1. Argonne National Laboratory (ANL), Argonne, IL (United States)
21. Wells L, Thurston S (2023) EVs@ scale next-gen profiles-fleet utilization 2023. No. ANL/TAPS-24/2. Argonne National Laboratory (ANL), Argonne, IL (United States)
22. Mishra DK, Zhang J, Lawler B, Castanier MP (2024) Parameter optimization for Li-ion battery module digital twin models. In: IECON 2024–50th annual conference of the IEEE industrial electronics society, pp 1–6. IEEE
23. Harizaj M, Bisha I, Basholli F (2023) IOT integration of electric vehicle charging infrastructure. Adv Eng Days (AED) 6:152–155
24. Zhang H, Lu F, Mi C (2020) An electric roadway system leveraging dynamic capacitive wireless charging: furthering the continuous charging of electric vehicles. IEEE Electr Mag 8(2):52–60
25. Langer E (2019) Liquid cooling for EV charging-What to know to keep electric vehicles on the go. CPC, Technical Guide 5011
26. Atallah RF, Assi CM, Fawaz W, Tushar MHK, Khabbaz MJ (2018) Optimal supercharge scheduling of electric vehicles: centralized versus decentralized methods. IEEE Trans Vehicul Technol 67(9):7896–7909
27. Anjana R, Kumar V (2024) Comprehensive review of electric vehicle charging infrastructure vulnerabilities and defense against malicious attacks. In: 2024 International conference on IoT, communication and automation technology (ICICAT), pp 1631–1638. IEEE

28. Mishra DK (2022) Quantifying the resilience enhancement of an active distribution system through multi-microgrid. University of Technology Sydney (Australia)
29. Kundu S, Giri AK, Kadiyan S, Singh S (2025) Power management and adaptive control method for a standalone distributed generation system under fault, DC-offset, and dynamic environments. Int J Ambient Energy 46(1). https://doi.org/10.1080/01430750.2025.2452245
30. Kundu S, Giri AK, Kadiyan S (2025) An adaptive mixed-step size normalized least means fourth control approach for stand-alone power generation system considering dynamic conditions. IEEE J Emerg Select Topics Power Electron 13(2):2422–2429. https://doi.org/10.1109/JESTPE.2024.3516878
31. Al-Musawi ANAA, Nema A, Femilinjana DL, Joshi A, Vimala Devi R (2024) Solar-driven wireless charging: innovations in electric vehicle energy systems. In: E3S web of conferences, vol 540, p 02025. EDP Sciences
32. Li M, Feng M, Luo D, Chen Z (2020) Fast charging Li-ion batteries for a new era of electric vehicles. Cell Rep Phys Sci 1(10)
33. Alberini A, Santoboni G (2025) Ownership and use of electric vehicles in the Washington, DC, metro area. Travel Behav Soc 39:100945
34. Das O, Zafar MH, Sanfilippo F, Rudra S, Kolhe ML (2024) Advancements in digital twin technology and machine learning for energy systems: a comprehensive review of applications in smart grids, renewable energy, and electric vehicle optimisation. Energy Convers Manag X:100715
35. Moniruzzaman, Yassine A, Shamim Hossain M (2023) Energizing charging services for next-generation consumers E-mobility with reinforcement learning and blockchain. IEEE Trans Consumer Electron 70(1):2269–2280
36. Xiaojian H, Zhang X, Xu, Liangcheng, Feng J, Luo S (2024) The battery swapping service network deployment problem: impact of driver range anxiety and impatience. Comput Ind Eng 192:110189
37. Qazi S, Khawaja BA, Alamri A, AlKassem A (2024) Fair energy trading in blockchain-inspired smart grid: technological barriers and future trends in the age of electric vehicles. World Electr Vehicle J 15(11):487
38. Kundu S, Singh M, Giri AK (2024) SPV-wind-BES-based islanded electrical supply system for remote applications with power quality enhancement. Electr Eng 106:279–294. https://doi.org/10.1007/s00202-023-01979-0
39. Abbasi MH, Arjmandzadeh Z, Zhang J, Krovi VN, Xu B, Mishra DK (2024) A coupled game theory and Lyapunov optimization approach to electric vehicle charging at fast charging stations. IEEE Trans Veh Technol 73(10):14224–14235
40. Kundu S, Singh M, Giri AK (2023) Adaptive control approach-based isolated microgrid system with alleviating power quality problems. Electr Power Components Syst 52(7):1219–1234. https://doi.org/10.1080/15325008.2023.2239222
41. Ghadi MJ, Mishra DK, Azizivahed A, Li L, Zhang J (2023) Mobile compressed air energy storage for active distribution systems. Int J Electr Power Energy Syst 154:109434
42. Verma SK, Kundu S, Singh S (2022) Design and techno-economic analysis of electric vehicle charging station for a railway metro station parking area. In: Mahajan V, Chowdhury A, Padhy NP, Lezama F (eds) Sustainable technology and advanced computing in electrical engineering. Lecture Notes in Electrical Engineering, vol 939. Springer, Singapore. https://doi.org/10.1007/978-981-19-4364-5_67
43. Kundu S, Singh M, Giri AK et al (2024) Robust and fast control approach for islanded microgrid system and EV charging station applications. Electr Eng. https://doi.org/10.1007/s00202-024-02291-1

Chapter 3
Hybrid Charging for Electrical Vehicle

Jigneshkumar P. Desai

Abstract Hybrid Electrical Vehicle (EV) charging systems utilize two or more power sources to ensure efficient and uninterrupted electric vehicle charging. This chapter specifically focuses on a tri-hybrid charging system, which integrates solar power, wind power, battery backup, and the electrical grid. The implementation of this system requires a combined solar-wind controller, Maximum Power Point Tracking (MPPT) technology, and a Battery Management System (BMS). The system operates intelligently by switching between power sources based on availability, weather conditions, and load demand, making it highly reliable, environmentally friendly, and cost-effective. The chapter begins with an overview of the system's working principles, followed by a detailed explanation of the modes of operation and various scenarios encountered during hybrid charging. In the final section, a case study using HOMER Pro simulation software is presented, demonstrating the practical application of the hybrid charging model. A tri-hybrid EV charging system was simulated using HOMER Pro with real solar and wind data. The system produced 690,866 kWh/year in which 97.5% from solar PV and only 2.5% from the grid. EV charging used 7.4% of the energy; 92.6% was exported due to surplus generation. V2G support at night improved grid resilience. The model shows high efficiency, low grid dependence, and better sustainability than conventional charging. The chapter concludes with a comparative analysis showing the performance and cost benefits of hybrid charging systems over conventional EV charging methods, highlighting its potential for significant advancements in sustainable transportation infrastructure.

Keywords Battery management system · Electric vehicle · Hybrid EV charging · Transportation

---

J. P. Desai (✉)
Department of Mechatronics Engineering & Department of Robotics and Automation Engineering, Parul Institute of Technology, Parul University, Vadodara 391760, India
e-mail: jpd.fetr@gmail.com; jigneshkumar.desai35873@paruluniversity.ac.in

S. Kundu et al. (eds.), *Hybrid Renewable Power Infrastructure for Sustainable Electric Vehicle Development*, Studies in Infrastructure and Control,
https://doi.org/10.1007/978-981-95-1323-9_3

1 Introduction

Electric vehicles (EVs) are at the forefront of the transition to sustainable transportation. However, their widespread adoption depends significantly on the availability, reliability, and efficiency of charging infrastructure. Conventional EV charging systems rely heavily on grid power, which may be unstable or limited, particularly in remote or underdeveloped areas. To address these challenges, hybrid charging systems are being developed, combining multiple energy sources to ensure consistent and eco-friendly power delivery. This chapter focuses on the design, operation, and simulation of a Tri-Hybrid EV charging system, which integrates solar power, wind power, battery backup, and grid support [1–3].

Recent studies have made significant contributions to the control and optimization of isolated and distributed generation systems integrating solar photovoltaic (SPV), wind energy conversion systems (WECS), and battery energy storage systems (BES). Their proposed adaptive control methods, including improved complex coefficient filter-based phase-locked loop (ICCF-PLL) [4], intelligent generalized maximum versoria criterion filtering (IGMVCF) [5], and mixed-step size normalized least means fourth (MSSNLMF) algorithms [6], enhance voltage and frequency regulation, harmonic suppression, and load balancing under dynamic renewable energy and load conditions. The use of bidirectional DC–DC converters (BDC) for power flow management between energy sources and loads is also emphasized [6, 9]. Furthermore, multiport battery charging and automated battery swapping concepts are introduced for efficient electric vehicle (EV) charging infrastructures, incorporating vehicle-to-home (V2H) and vehicle-to-vehicle (V2V) power transfer [5, 7]. Experimental validations demonstrate compliance with Institute of Electrical and Electronics Engineers (IEEE) standards, maintaining low total harmonic distortion (THD) and ensuring high power quality (PQ) [4–9]. Collectively, these works advance sustainable, robust microgrid systems suitable for remote and islanded applications.

2 Concept of Hybrid EV Charging

Hybrid EV charging systems utilize two or more energy sources in combination to provide uninterrupted charging capability. The tri-hybrid model discussed in this chapter uses solar energy during daylight hours, wind energy during nighttime or breezy conditions, and draws from the grid when renewable sources are insufficient. Additionally, a battery energy storage system and EV ensures buffer energy supply during peak load times or sudden power drops [10]. The implementation of a tri-hybrid EV charging system requires the following key components:

Solar Panels: Harvest solar energy during daylight hours. Figure 1 shows the solar farm. Each limb has 48 panels out of which 24 are in series and such 2 modules are in

Fig. 1 Photovoltaic system solar panels. Image by mrganso, used under the Pixabay Content License [11]

parallel. Table 1 shows comparison between monocrystal, polycrystal, and thin-film solar panels.

Wind Turbines: Generate power from wind energy, especially useful at night. Figure 2 shows the horizontal axis wind turbine plant. Table 2 compares horizontal and vertical axis wind turbines.

Battery Storage System: Stores excess power and delivers it when needed. As shown in Table 3, battery storage system for less than 50 kW is lithium-ion type and having 3000 to 7000 cycle life.

Combined Solar-Wind Controller: Manages input from both renewable sources.

Maximum Power Point Tracking (MPPT): Optimizes energy extraction from solar and wind systems. Common MPPT algorithms are as follows:

(1) Perturb and Observe (P&O): This method slightly perturbs the operating voltage and observes the change in power. If power increases, the perturbation continues in the same direction; if it decreases, the direction is reversed. It's simple but can cause oscillations around the MPP.
(2) Incremental Conductance: This technique calculates the derivative of power with respect to voltage (dP/dV) and adjusts the operating point accordingly. It can track the MPP more accurately under rapidly changing conditions compared to P&O.

Table 1 Comparison of different types of solar panels

Parameter	Monocrystalline	Polycrystalline	Thin film
Efficiency	18–24% (highest)	14–17%	Lower than crystalline panels
Material	Single-crystal silicon	Multiple silicon fragments	Amorphous silicon/ CdTe/CIGS
Appearance	Sleek black	Blue	Thin, flexible
Space requirement	Compact (best for small roofs)	Needs more space	Highest space needed
Performance	Best in heat & low light	Moderate in heat	Good in low light
Lifespan	25+ years	20–25 years	10–15 years
Cost (per panel)	₹8,000–₹13,500	₹5,000–₹10,000	₹1,000–₹7,000 (varies by material)
Best for	Homes with limited space	Budget-conscious buyers	Commercial/portable use
Top brands	Loom Solar, SolarSquare, Freyr Energy	Bluebird Solar, Waaree	First Solar, Tata Power Solar

Fig. 2 Growian wind turbine. Image by Pixabay, used under the Pixabay Content License [12]

Table 2 Comparison of different types of wind turbines

Category	Horizontal-axis wind turbines (HAWT)	Vertical-axis wind turbines (VAWT)
Design	Blades rotate around a horizontal axis (traditional windmill design)	Blades rotate around a vertical axis (eggbeater-like design)
Prevalence	Most common type used today	Less commonly used
Efficiency	Generally, more efficient	Typically, less efficient than HAWTs
Wind direction	Needs to face wind direction (requires yaw mechanism)	Can capture wind from any direction (no need to face wind)
Installation height	Typically mounted on tall towers	Can be installed closer to the ground
Applications	– Residential systems	– Urban environments
	– Utility-scale wind farms	– Small-scale installations

Table 3 Specification of BESS

Parameter	Small microgrids (<50 kW)
Battery storage	
–Type	Lithium-ion
–Energy density	150–200 Wh/kg
–Cycle life	3,000–7,000 cycles
–DoD	Up to 80%
Inverters	
–Type	Single-phase (3 kW–2 MW)
–Efficiency	95–98%
–Key features	MPPT, grid-tie, anti-islanding

(3) Constant Voltage (CV): This approach maintains the panel voltage at a fixed percentage of the open-circuit voltage, typically around 76%. It's less accurate but simpler to implement.

Battery Management System (BMS): Controls charging/discharging cycles and maintains battery health. In renewable energy applications, such as solar- or wind-powered microgrids, a BMS is vital for

(1) Maximizing battery lifespan by preventing harmful operating conditions.
(2) Ensuring efficient energy storage and retrieval, aligning with variable energy generation patterns.
(3) Providing real-time data for system optimization and maintenance planning.

Grid Connection: It refers to the process of linking a local power generation system, such as solar panels, wind turbines, or a hybrid energy system, to the public electricity grid. This integration allows for the exchange of electricity between the local system and the grid, creating a two-way flow of energy. Provides a backup source of energy when renewable sources are unavailable or insufficient.

Smart Energy Management System: A Smart Energy Management System (SEMS) is a transformative approach to energy management, combining renewable energy sources, automation, and advanced technologies to improve efficiency, reduce costs, and support sustainability. Whether in homes, commercial buildings, or industrial applications, SEMS offers tremendous potential to revolutionize energy consumption and contribute to a more sustainable future. Automates energy switching and load distribution based on real-time conditions [13].

Electrical vehicle: There are different types of electric vehicles based on their power sources and charging methods [14]:

(1) Battery Electric Vehicles (BEVs): These are fully electric vehicles that rely solely on battery power for propulsion. They do not have a gasoline engine, and they must be recharged via an electric outlet or charging station. Example: Tesla Model 3, Nissan Leaf, and Chevrolet Bolt EV.
(2) Plug-in Hybrid Electric Vehicles (PHEVs): These vehicles have both an electric motor and a gasoline engine. The electric motor can be charged via an outlet, and when the battery runs out, the vehicle switches to the gasoline engine. Example: Toyota Prius Plug-in, Mitsubishi Outlander PHEV.
(3) Hybrid Electric Vehicles (HEVs): HEVs combine an internal combustion engine with an electric motor, but they do not require external charging. The battery is charged through regenerative braking and the engine itself. Example: Toyota Prius, Honda Insight.
(4) Fuel Cell Electric Vehicles (FCEVs): These vehicles use hydrogen fuel cells to generate electricity that powers the electric motor. They are refueled with hydrogen instead of being plugged into a charger. Example: Toyota Mirai, Hyundai Nexo.

3 Working Principle

The tri-hybrid system operates dynamically, choosing the most suitable energy source or combination of sources based on real-time parameters such as solar radiation, wind speed, energy demand, and battery charge level. During the day, the system prioritizes solar energy; at night or in cloudy conditions, it switches to wind power. In situations where both solar and wind are inadequate, the system draws power from the grid. The battery bank acts as a buffer, storing excess energy and supplying it when demand spikes or renewables are momentarily inactive. Consider the system as shown in Table 4 having three sources such as solar PV, wind and storage system. Grid can be used in extreme cases that will be explained later in this chapter. For this system, daily operation is explained in next subsection.

Table 4 Tri-hybrid charging system components

Component	Specification	Purpose
Solar PV	50 kW	Primary daytime power source
Wind turbine	20 kW (horizontal axis)	Complements solar, especially effective at night
Battery bank	200 kWh Li-ion	Stores excess renewable energy, manages peak loads
Grid supply	3-phase, 50 kVA	Acts as a backup during low renewable availability

3.1 Typical Daily Operation

The Techno Green Campus considered as load in order to explained daily operations, equipped with a 50 kW solar plant, a 20 kW wind turbine, a 200 kWh battery bank, and a 50 kVA grid connection, now includes electric vehicle charging stations that support faculty, staff, and student vehicles. EVs charge primarily during the day and can potentially support Vehicle-to-Grid (V2G) operations during peak demand or low renewable availability [15]. Table 5 shows major time event of tri-hybrid charging systems with Electrical Vehicle (EV). Figure 3 plots the energy curve and contribution of each source during the daily operation. It also shows the total energy required.

With the addition of Electric Vehicles (EVs), the smart campus tri-hybrid system at Techno Green Campus transforms into a more dynamic and responsive energy ecosystem. The plot of Fig. 1 demonstrates the campus's self-sufficiency during the day, where solar and wind meet all energy demands, with the battery storage supporting when needed. However, at night, the campus relies on the grid when renewable generation is low, but EVs can potentially help reduce grid dependency via Vehicle-to-Grid (V2G).

As shown in Fig. 3, the morning at 9:00 AM, the solar panels begin producing near-peak power (~45 kW) while wind contributes modestly (~3 kW). As faculty and staff arrive, EVs begin charging, adding a new 10 kW load. The 30 kW required by the campus infrastructure plus the 10 kW EV charging is fully met by solar, and any excess energy (~2 kW) starts charging the battery.

By 1:00 PM, during peak solar conditions, the system hits its stride. Solar output climbs to 50 kW and wind reaches 10 kW. With around 15 kW of EV load and 35 kW of campus load, the total 50 kW demand is met entirely by renewables. The battery

Table 5 Daily time series operation of tri-hybrid charging system

Time	Solar	Wind	Battery	EV role	Grid
9:00 AM	Available	Available	Charge	Starts charging	Not connected
1:00 PM	Available	Available	Full	Full charging	Not connected
7:00 PM	Not available	Available	Discharge	Minimal charging	Not connected
11:00 PM	Not available	Available	Low	Idle/V2G	Connected

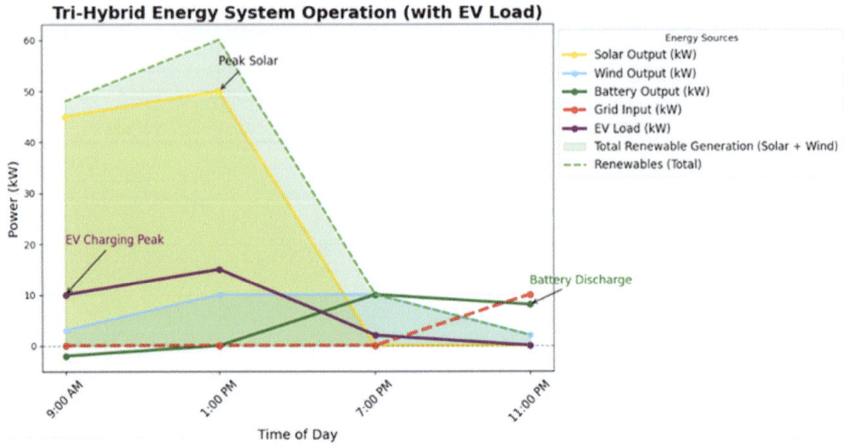

Fig. 3 Tri-hybrid charging system daily operating state

is fully charged and remains in standby. No grid energy is needed, highlighting a state of complete energy self-sufficiency, even with the added EV demand.

In the evening at 7:00 PM, solar production ceases, but wind output remains steady at around 10 kW. EV charging is reduced or paused since most vehicles are already charged. The remaining 20 kW campus load is partially supplied by wind, while the battery discharges to bridge the gap. No grid input is required during this period, showcasing the smart system's ability to seamlessly handle transitions between sources.

By 11:00 PM, renewable generation dips significantly—solar is absent and wind falls to just 2 kW. The battery's state of charge drops to around 30% after evening discharge. To avoid deep discharge and preserve battery life, the system draws about 10–12 kW from the grid to meet the remaining load. If Vehicle-to-Grid (V2G) is enabled, pre-authorized EVs can temporarily supply 1–2 kW each, reducing grid reliance even further.

4 Modes of Operation

The tri-hybrid charging system for Electric Vehicles (EVs) integrated into a microgrid begins with the Photovoltaic (PV) system, which consists of a solar panel and a DC source such as a battery backup or capacitor as depicted in Fig. 4. This system generates DC voltage (V_{PV}) and current (I_{PV}) based on the available solar irradiation. The generated power is fed into a High Gain ZETA-SEPIC Converter, which performs DC–DC conversion and boosts the voltage to the required level.

The High Gain ZETA-SEPIC Converter combines the advantages of both ZETA and SEPIC converter topologies to achieve high voltage gain and effective control. It

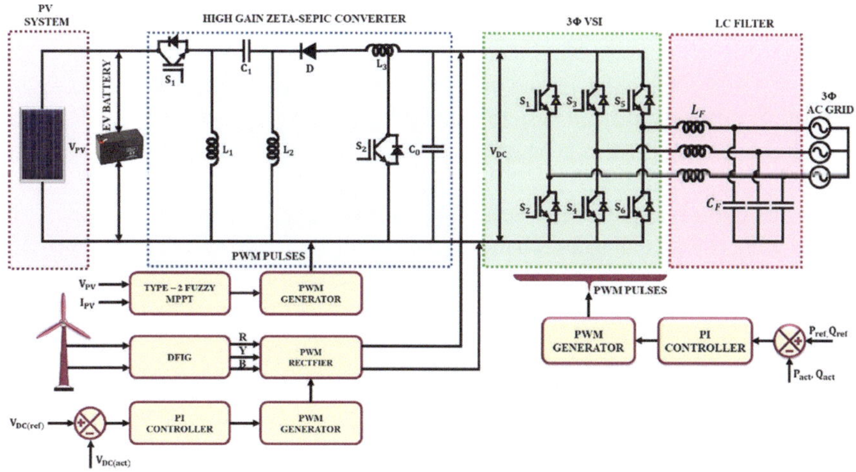

Fig. 4 Tri-hybrid charging system circuit diagram [16]

includes switches S_1 and S_2, inductors L_1 and L_2, capacitors C_1 and C_0, and a diode D. This converter boosts the voltage from the PV system and transfers the output to the DC link capacitor of the three-phase Voltage Source Inverter (VSI). The operation of this converter is regulated by a Type-2 Fuzzy Logic-based Maximum Power Point Tracking (MPPT) controller to ensure that the PV system operates at its maximum efficiency.

In addition to solar energy, the system integrates a wind energy system based on a Doubly Fed Induction Generator (DFIG). The DFIG is interfaced with the DC link through a rotor-side Pulse Width Modulation (PWM) rectifier and a PWM generator. The voltage output from the DFIG is regulated by a Proportional–Integral (PI) controller, which ensures the DC voltage (V_{DC}) matches a desired reference value ($V_{DC(ref)}$), thus enabling stable voltage injection from the wind system into the DC bus. The combined DC output from both PV and wind sources is supplied to the three-phase voltage source inverter (3-Φ VSI). This inverter, comprising six IGBT switches (S_1 to S_6), converts the DC voltage into three-phase AC voltage. The switching of these IGBTs is controlled using PWM signals generated by the system's controller. To ensure a smooth AC output with minimal harmonics, the inverter output is passed through an LC filter, which includes inductors (L_F) and capacitors (C_F). This filter conditions the waveform and makes it suitable for integration into the three-phase AC grid, which represents the distribution interface for supplying power to EV charging stations or other local loads. The grid connection ensures reliable power delivery and voltage stability.

The system's performance is managed using various controllers and pulse generators. The MPPT controller uses type-2 fuzzy logic to maximize power extraction from the PV source. PI controllers regulate the DC link voltage and manage active (P_{ref}) and reactive (Q_{ref}) power delivery, while PWM generators produce the necessary gate signals for the operation of converters and inverters.

In terms of power flow coordination, the solar energy is converted and boosted via the ZETA-SEPIC converter and then passed through the inverter to the grid. Wind energy is rectified, regulated, and similarly delivered to the grid via the common DC link. If a battery storage system is present, it can be connected to the DC link for charging during surplus generation or discharging during peak demand, thereby supporting the overall load.

This tri-hybrid system offers several key advantages. The hybrid integration of PV and wind energy enhances reliability and sustainability. High-efficiency DC–DC conversion is achieved through the advanced converter topology. Grid synchronization and voltage stability are maintained using the LC filter and PI-based inverter control. Additionally, smart control strategies like type-2 fuzzy MPPT improve the system's dynamic performance and energy utilization. Different modes during the day are as follows:

(1) Solar-Dominant Mode: On sunny days, solar energy meets most or all of the demand.
(2) Wind-Dominant Mode: On windy nights, wind energy becomes the primary source.
(3) Battery Support Mode: Battery discharges to supplement insufficient renewable supply.
(4) Grid Support Mode: Grid compensates when solar, wind, and battery are unable to meet the load.
(5) Mixed Mode: Simultaneous use of solar, wind, and grid power to maintain load.

5 Case Study: HOMER Pro Simulation

To evaluate the feasibility and performance of the tri-hybrid charging model, a simulation was conducted using HOMER Pro. The model of Fig. 5 designed using real-world solar irradiance and wind speed data, incorporating system costs, load profiles, and energy price inputs. Various configurations were tested to determine the most cost-effective and efficient design. The HOMER Pro system parameters are as shown in Table 6.

A comprehensive simulation was performed using the HOMER Pro software to evaluate the techno-economic feasibility of a hybrid energy system integrated with electric vehicle (EV) charging infrastructure [17]. The proposed system architecture of Fig. 5 includes a large-scale solar PV array, pumped hydro-storage, and grid connectivity.

The plotted results of Fig. 6 provide a concise visualization of the system's energy dynamics. The total energy production from the system amounts to 6,90,866 kWh/year, of which 97.5% is contributed by the solar PV array, showcasing high solar penetration. Only 2.5% of the energy is purchased from the grid, highlighting the systems near self-sufficiency. In terms of energy consumption, 7.4% of the generated energy is utilized to meet the direct EV charging load. A substantial 92.6% is exported

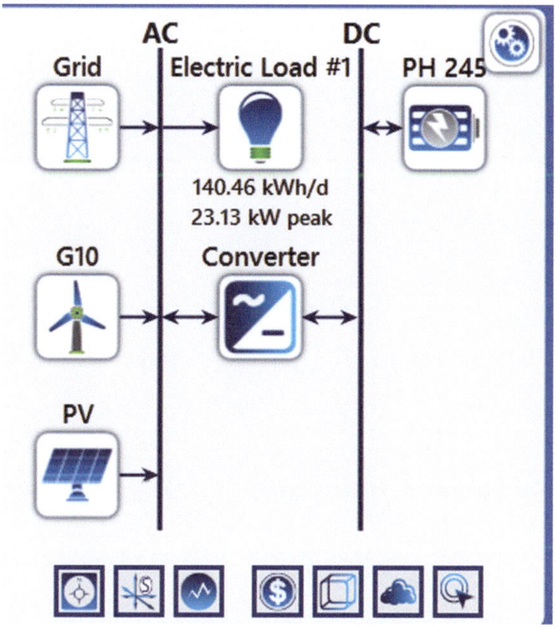

Fig. 5 HOMER Pro system design

back to the grid, facilitated by favorable sellback rates and minimal local demand, particularly during peak generation hours.

The EV charging schedule is aligned with solar generation hours (9 AM to 2 PM), ensuring direct utilization of renewable energy and minimizing reliance on storage or the grid. The Vehicle-to-Grid (V2G) feature is enabled at night, allowing EVs to discharge back into the system when renewable availability is low, effectively supporting grid and storage shortfalls.

This simulation demonstrates the feasibility of a solar-dominant hybrid system with grid interaction and EV integration. The results strongly support the economic and operational viability of decentralized renewable energy systems in urban Indian contexts.

6 Comparison with Conventional Charging

Compared to traditional grid-only EV charging systems, the tri-hybrid model offers several advantages:

(1) Sustainability: Utilizes clean energy sources, reducing carbon footprint.
(2) Resilience: Continues operation even during grid outages or fluctuations.
(3) Cost Efficiency: Reduces electricity bills by minimizing grid dependency.
(4) Scalability: Can be customized for urban, semi-urban, and rural deployment.

Table 6 HOMER Pro system specification

Component	Parameter	Value	Unit	Description
Solar PV array	Rated capacity	50	kW	Maximum output during peak sunlight hours
	Derating factor	0.9	–	Accounts for dust, wiring, temperature losses
	Tracking mode	Fixed tilt	–	No tracking system used
	Capital cost	1,000	$/kW	Typical market value
	Lifetime	25	Years	Standard PV lifespan
Wind turbine	Rated capacity	20	kW	Combined capacity of small turbines
	Hub height	25	meters	Medium tower height
	Cut-in wind speed	3	m/s	Minimum wind speed for generation
	Rated wind speed	12	m/s	Optimal performance wind speed
	Capital cost	2,000	$/kW	Installed cost per kW
	Lifetime	20	Years	Typical for small-scale wind turbines
Battery bank	Nominal capacity	200	kWh	Total usable capacity
	Round-trip efficiency	90	%	Energy retention rate
	Max depth of discharge	80	%	To extend battery life
	Min. state of charge	20	%	Lower limit for grid fallback triggering
	Capital cost	400	$/kWh	Li-ion battery estimate
	Lifetime	10	Years	HOMER default for lithium-ion

(continued)

Table 6 (continued)

Component	Parameter	Value	Unit	Description
Grid connection	Maximum capacity	50	kVA	Backup capacity during low renewable availability
	Grid availability	100	%	Assume full grid availability
	Grid tariff (Buy)	0.1	$/kWh	Cost for buying power from grid
	Grid sellback rate	0.05	$/kWh	Price received when selling excess energy
EV load profile	Charging power range	10–15 (peak), ~2 (night)	kW	Based on operational schedule
	Time of use	9 AM–2 PM	Hours	EVs mainly charge during solar hours
	Bidirectional V2G capability	Enabled at night	–	Provides support during grid or battery shortfall
	V2G discharge power	1–2	kW per EV	As per narrative description

7 Conclusion

The tri-hybrid EV charging system presents a robust and sustainable approach to addressing the growing energy demands of electric mobility. By integrating solar PV, wind energy, battery storage, and grid connectivity, the system achieves a balanced mix of reliability, cost-effectiveness, and environmental sustainability. The HOMER Pro simulation results affirm the technical and economic feasibility of this model, demonstrating high solar energy penetration (97.5%) and minimal grid reliance (2.5%), with effective utilization of vehicle-to-grid (V2G) capabilities.

This approach not only reduces dependence on conventional grid power but also ensures resilience during supply fluctuations and peak demand periods. Its adaptability to various geographic and demographic contexts such as urban, semi-urban, and rural such as further enhances its potential for large-scale deployment.

Looking ahead, integrating smart control strategies such as artificial intelligence, demand forecasting, and adaptive energy management algorithms could significantly enhance system performance and user responsiveness. The tri-hybrid architecture thus holds promise as a scalable solution for clean, reliable, and intelligent EV charging infrastructure in energy-transition economies like India.

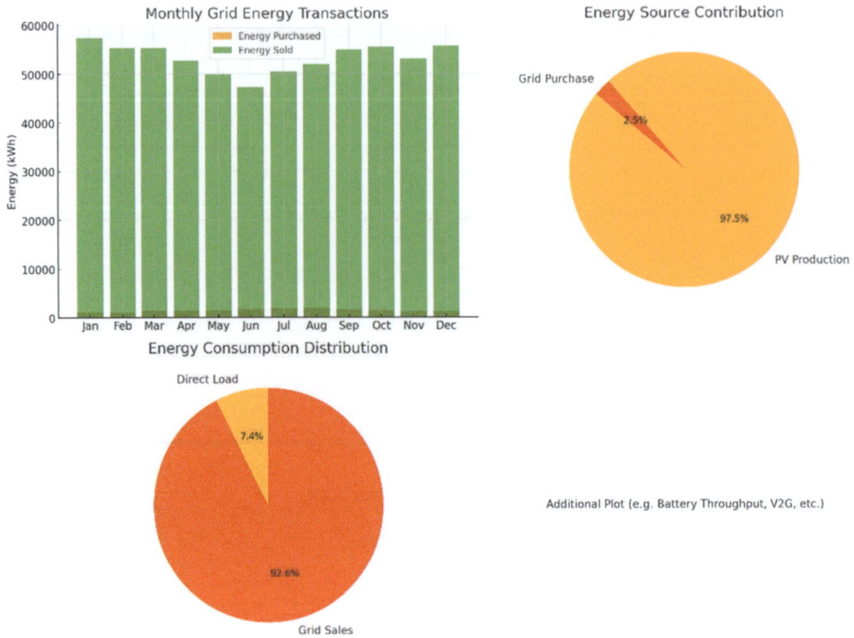

Fig. 6 Energy utilization chart

Acknowledgements NA

Competing Interests NA

Ethics Approval NA

References

1. Desai JP (2025) Smart grid and energy management. In: Giri AK, Singh M (eds) Electric vehicle charging infrastructures and its challenges. Studies in infrastructure and control. Springer, Singapore. https://doi.org/10.1007/978-981-96-0361-9_6
2. Desai JP, Dubey R, Desai A (2024) Modeling and implementation of autorecloser in radial power system and microgrid. In: 2024 IEEE 3rd international conference on electrical power and energy systems (ICEPES). Bhopal, India, pp 1–5. https://doi.org/10.1109/ICEPES60647.2024.10653500
3. Majid MA, Kumar J, R C, Ahmed A (2024) Advances in electric vehicles for a self-reliant energy ecosystem and powering a sustainable future in India. e-Prime-Adv Electr Eng Electron Energy 10
4. Kundu S, Singh M. Giri AK (2023) Synchronization and control of WECS-SPV-BSS-based distributed generation system using ICCF-PLL control approach. Electric Power Syst Res. https://doi.org/10.1016/j.epsr.2023.109919
5. Kundu S, Giri AK, Singh M, Singh S (2025) Home electric vehicle charging infrastructures and its challenges: effective multiport battery charging facility, operation and control in hybrid isolated microgrid system. In: Studies in infrastructure and control

6. Kundu S, Giri AK, Kadiyan S (2024) An adaptive mixed-step size normalized least means fourth control approach for stand-alone power generation system considering dynamic conditions. IEEE J Emerg Select Topics Power Electron 13(2):2422 2429. https://doi.org/10.1109/JESTPE.2024.3516878
7. Kundu S, Singh M, Giri AK (2024) Robust and fast control approach for islanded microgrid system and EV charging station applications. Electrical Engineering
8. Kundu S, Singh M, Giri AK (2024) SPV-wind-BES-based islanded electrical supply system for remote applications with power quality enhancement. Electr Eng 106:279–294
9. Kundu S, Singh M, Giri AK (2023) Adaptive control approach-based isolated microgrid system with alleviating power quality problesms. J Electr Eng Technol 1219–1234. https://doi.org/10.1080/15325008.2023.2239222
10. Singh P, Tripathi S (2023) Solar powered hybrid charging station for electrical vehicle. ResearchGate. https://doi.org/10.13140/RG.2.2.11216.69127
11. mrganso (2017) Photovoltaic system solar [Photograph]. Pixabay. https://pixabay.com/photos/photovoltaic-system-solar-2742302/
12. Pixabay (2011) Growian wind turbine [Photograph]. Pixabay. https://pixabay.com/photos/growian-wind-wind-turbine-11205/
13. Baba AO, Liu G, Chen X (2020) Classification and evaluation review of maximum power point tracking methods. Susta Futures 2:100020. https://doi.org/10.1016/j.sftr.2020.100020
14. G K, Ediga P, S A, P A, T S, Mittal A, Habelalmateen MI (2024) Smart energy management: real-time prediction and optimization for IoT-enabled smart homes. Cogent Eng 11(1). https://doi.org/10.1080/23311916.2024.2390674
15. Discover Engineering (2025) Electric vehicles. Discover Engineering. https://www.discoverengineering.org/electric-vehicles/
16. Kumar RT, Rajan CCA (2023) Integration of hybrid PV-wind system for electric vehicle charging: towards a sustainable future. e-Prime Adv Electr Eng Electron Energy 6:100347. https://doi.org/10.1016/j.prime.2023.100347
17. HOMER Energy (2020) HOMER Pro® (Version 3.11.2) [Computer software]. HOMER Energy by UL. https://www.homerenergy.com/products/pro/index.html

Chapter 4
Machine Learning-Based Grid-Interactive Chargers for Optimizing Power Quality Under Non-grid Conditions

Gaurav Yadav, Sudhanshu Mittal, and Sombir Kundu

Abstract Bi-directional power flow made possible by vehicle-to-grid (V2G) technology in grid-interactive electric vehicle (EV) chargers lets EVs not only consume but also provide electricity from the grid. This capacity improves grid stability and energy management and provides financial gains, such as selling electricity during peak demand. Although advancements in power electronics and converter topologies have increased system efficiency, it is still difficult to ensure power quality under less-than-ideal grid conditions. The study explores the use of machine learning (ML) in conjunction with the decision tree learning approach to replace traditional PI controllers, therefore reducing total harmonic distortion (THD) and improving grid stability. Finally, a 1.1 kW MATLAB/Simulink simulation supports the efficiency of the proposed system.

Keywords Machine learning · Wide voltage · EV charger

1 Introduction

The Grid-Interactive Electric Vehicle Charger (GIEVC) mainly interacts with the grid in terms of power exchange such that power flows from the grid to the vehicle (G2V) and back to the grid from the vehicle (V2G). This bi-directional flow of power benefits in most of the cases. Such that while giving power back to the grid it can be

G. Yadav
Electrical Engineering, Delhi Technological University, Delhi, India
e-mail: gauravyadav_2k21phdee17@dtu.ac.in

S. Mittal (✉)
Thapar Institute of Engineering and Technology, Patiala, Punjab, India
e-mail: sudhanshu.mittal@thapar.edu

S. Kundu
MMEC, Maharishi Markandeshwar (Deemed to be University), Ambala, Haryana, India

S. Kundu et al. (eds.), *Hybrid Renewable Power Infrastructure for Sustainable Electric Vehicle Development*, Studies in Infrastructure and Control,
https://doi.org/10.1007/978-981-95-1323-9_4

advantageous in terms of selling electricity when there is the peak demand [1–5]. In addition, these techniques like V2G also help in frequency support to maintain the grid. Further, power electronics advancements like wide-bandgap semiconductors help in increasing the efficacy of power flow between EVs and the grid [6–9].

Further, as discussed above about Vehicle-to-Grid (V2G) technology, the main query comes from the power balance at the point of common coupling (PCC). The non-linear load connected in the shunt at the PCC is a representation of residential load or emergency load [10]. The control scheme used to generate the pulses for the AC/DC converter should take care regarding the power balance at the PCC. More specifically talking about the DC-link voltage control, which is responsible for the reactive power demand created by the non-linear load [11–13].

The seamless control of the RSMLI-based AC/DC converter in grid-connected EV systems [14–16]. It uses many algorithms, such that the most commonly used control techniques are the second-order generalized integrator (SOGI) [17], ARBFNN [18], and LLMF [19]. For G2V/V2G operation, the SOGI algorithm is intended to control these converters, and its DC-link is connected as a medium between a DC–DC converter. Further, a filter is connected after the DC/DC converter to filter out the output current to charge the battery [20].

To control the DC-link bus voltage, a machine learning-based algorithm is used. The advantage of ML-based techniques is to optimize the efficiency and reliability of EV charging systems [21]. These systems often involve complex interactions between various components, such as converters, energy storage units, and the grid. Machine learning techniques, including decision trees, neural networks, fuzzy logic, and reinforcement learning, are employed to manage these interactions effectively [22]. This response explores the application of machine learning in DC-link control for EV charging, focusing on different control strategies and their benefits [23–26].

This refers to the structure of the paper. In Sect. 1, the recommended work and literature review are explained. In Sect. 2, the details of the system are covered. In Sect. 3, the designed controller is covered. Section 4 covers the results and discussion, whereas Sect. 5 summarizes the conclusion.

2 System Configuration

The On-Board Bi-Directional Charger (OBBC) consists of two stages as depicted in Fig. 1. The first stage is the Grid Converter Voltage Source Converter (VSC) which is a full H-bridge made up of four IGBTs ($\boldsymbol{S}_1, \boldsymbol{S}_2, \boldsymbol{S}_3, \boldsymbol{S}_4$) and the second stage is Bipolar Dual-Directional DC (BDDC) (also called synchronous buck-boost converter) which consists of half two IGBTs ($\boldsymbol{S}_5, \boldsymbol{S}_6$). These IGBT switches are capable of bi-directional flow. Further, the source inductor $\boldsymbol{L_g}$ is connected in series with the single-phase voltage ($\boldsymbol{v_g}$), non-linear load and VSC via compensating inductor $\boldsymbol{L_f}$ are connected at the point of common coupling (PCC). A high capacitor ($\boldsymbol{C_{DC}}$) is connected at the output of the VSC, forming the DC-link. Also, filter

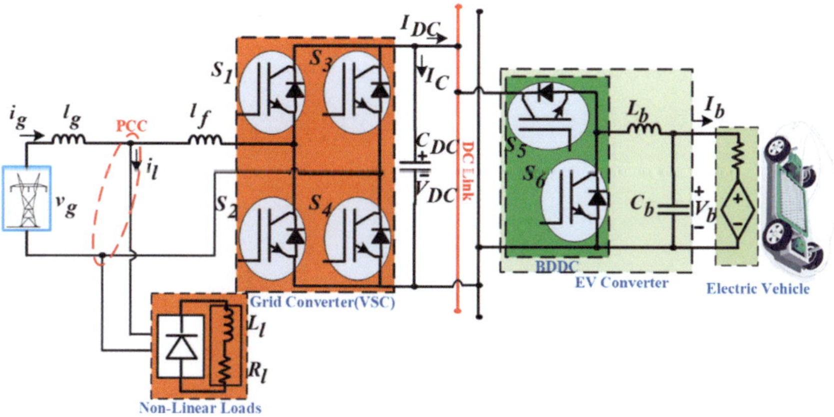

Fig. 1 Circuit diagram of OBBC

inductor $\boldsymbol{L_b}$ and capacitor $\boldsymbol{C_b}$ are connected after BDDC formed EV converter. At last EV battery is connected having $\boldsymbol{V_b}$ voltage across it.

3 Design of the Controller

In this section, the switching condition of VSC and the strategy to control BDDC are discussed in detail.

3.1 VSC Switching Strategy

From Fig. 2, it can be observed that the Cascaded Non-Identical Second-Order Generalized Integrator (CNISOGI) filter is connected to calculate the fundamental components $v_{gf(\alpha\beta)}$ and $i_{lf(\alpha\beta)}$ from the v_g and load current i_l. These fundamental components are further taken to calculate the v_{gf} as follows:

$$v_{gf} = \sqrt{\left(v_{gf\alpha}^2 + v_{gf\beta}^2\right)} \tag{1}$$

$$v_{gp} = \frac{(v_{gf\alpha})}{((v_{gf\alpha})^2 + (v_{gf\beta})^2)^{0.5}} \tag{2}$$

The VSC aims to regulate the DC-link voltage and to support the bi-directional power flow, a ML-DC-Link Voltage control (ML-DLVC) is used. Moreover, the difference of V_{DC} and V_{DC}^* is fed to ML-DLVC, to get the fundamental current

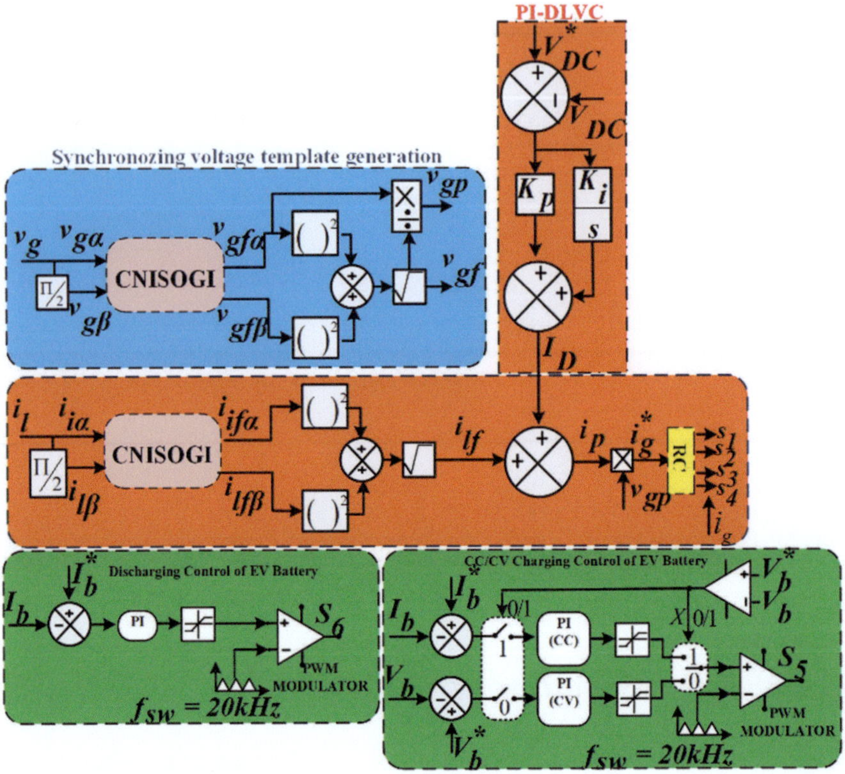

Fig. 2 Control architecture of OBBC

component I_D. Decision trees (DT) are a widely used analytical model in machine learning and statistical analysis, helping in decision-making and predicting outcomes based on given data.

In DT analysis, a fitness function evaluates how effective the splits are, using criteria like Gini impurity or information gain. Consider a training set L, where L = $\{(x_1, c_1),(x_2, c_2),\ldots,(x_j, c_j)\}$, with $(x_1, x_2, \ldots, x_i)$ being input feature vectors and $(c_1, c_2, \ldots, c_j)$ as corresponding class labels. The flowchart of the decision tree is depicted in Fig. 3. Here, x_i is a vector of input attributes, and the conditions for splitting are determined based on one of these attributes. Let p_i denote the probability that a random sample falls into class c_i; then p_i can be calculated as follows:

$$p_i = \frac{c_i}{L} \tag{3}$$

Information Gain

Information gain measures how much a feature f reduces uncertainty or entropy in a dataset before and after it's divided based on f. It indicates the effectiveness of f in

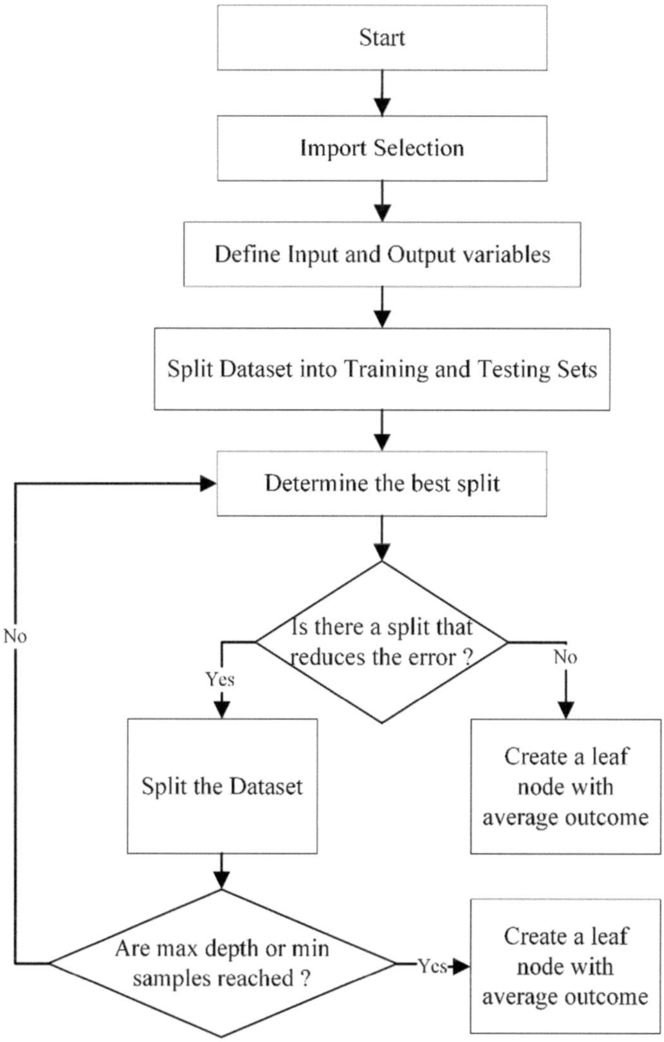

Fig. 3 Flowchart for DT algorithm

separating the dataset L into subsets $|L^v|$, where each subset represents data with a distinct value of f. A higher information gain suggests a feature is better at improving classification clarity. In Fig. 4, the prediction error plot and also the validation of the V_{DC}. The best feature for splitting is the one with the highest information gain, implying it offers the most significant reduction in uncertainty.

$$IG(L, f) = Entropy(L) - \sum_{v=1}^{v} \frac{|L^v|}{L} Entropy(L^v) \tag{4}$$

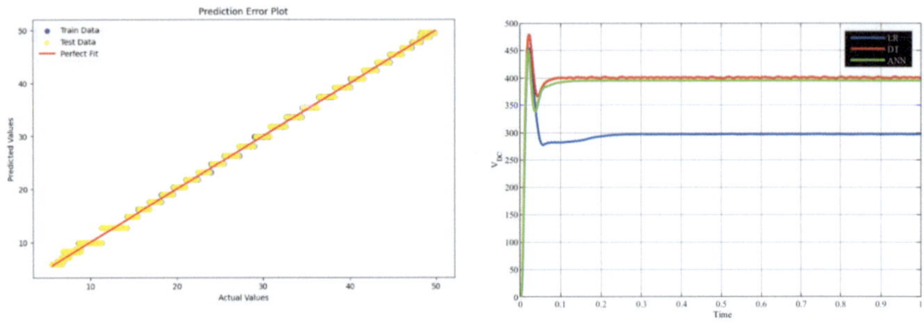

Fig. 4 Prediction error plot and validation for ML-DLV

The fitness function is calculated concerning MSE, MAE, and R2 square.

$$MSE = \frac{1}{n}\sum_{i \in n}\left(Y_i - \tilde{Y}_t\right)^2 \tag{5}$$

$$MSE = \frac{1}{n}\sum_{i \in n}\left|Y_i - \tilde{Y}_t\right| \tag{6}$$

$$R^2 = 1 - \frac{\sum_{i \in n}\left(Y_i - \tilde{Y}_t\right)^2}{\sum_{i \in n}\left(Y_i - \tilde{Y}\right)^2} \tag{7}$$

It is now necessary to include the DC-link loss factor I_D as well as an essential load current factor i_{lf} in order to obtain the overall active current reference (I_p).

$$R^2 = 1 - \frac{\sum_{i \in n}\left(Y_i - \tilde{Y}_t\right)^2}{\sum_{i \in n}\left(Y_i - \tilde{Y}\right)^2} \tag{8}$$

As per (9), multiplying the active current reference I_p with the unit voltage template v_{gp} yields reference sinusoidal current for active power.

$$i_g^* = I_p * v_{gp} \tag{9}$$

Ultimately, these intended reference values can be determined by comparing them with the actual grid current (i_g), which is then sent to the Hysteresis Current Control (HCC) to calculate gating signals for the grid-side converter's IGBT switches.

3.2 BDDC Control Strategy

The main objective is the controlled BDDC for an efficient battery charging system in the event of grid outages. Charging is managed using constant current (CC) and constant voltage (CV) methods. The battery is first charged using CC to regulate the charging process. When the battery voltage exceeds permitted limits, CV charging is utilized. This method allows for efficient charging despite exceeding the charger's voltage and current restrictions. After that, with the DC-link voltage nearly at its rated level, the battery reference current has to be changed to accommodate both CC and CV charging. Because BDDC is bi-directional, it can operate in both charging and draining modes. As a result, the controller needs to adhere to the battery's reference current (I_b^*), which can be adjusted to satisfy particular needs. Figure 2 shows the whole BDDC controller structure.

4 Result and Discussion

To prove the efficacy of the system, the training performance and testing performance are given in Tables 1 and 2. So, the MATLAB/Simulink is used to develop and simulate the system methodically utilizing different components from the Simscape package. The simulation's results are shown for two different periods. Figure 5 shows the controller's efficacy in both G2V and V2G modes during the first several hours of operation. To ensure UPF grid operation and fulfill the reactive power requirements of neighboring loads, the charger initially runs in G2V mode. At 0.5 s, the mode switches to V2G. Even in situations when there is a demand for reactive power at the PCC (point of common coupling), the UPF is observed in the grid in V2G mode to support reactive power via VSC. As per the present grid profile, communication between the grid and active power is limited.

Table 1 Training performance

Model number	Model type	MSE	R squared	MAE
1	Tree	0.0039	0.999	0.03
2	Neural network	0.076	0.997	0.16
3	Linear regression	0.84	0.97	0.55

Table 2 Testing performance

Model number	Model type	MSE	R squared	MAE
1	Tree	0.003	0.999	0.0379
2	Neural network	0.171	0.994	0.29
3	Linear regression	0.844	0.974	0.557

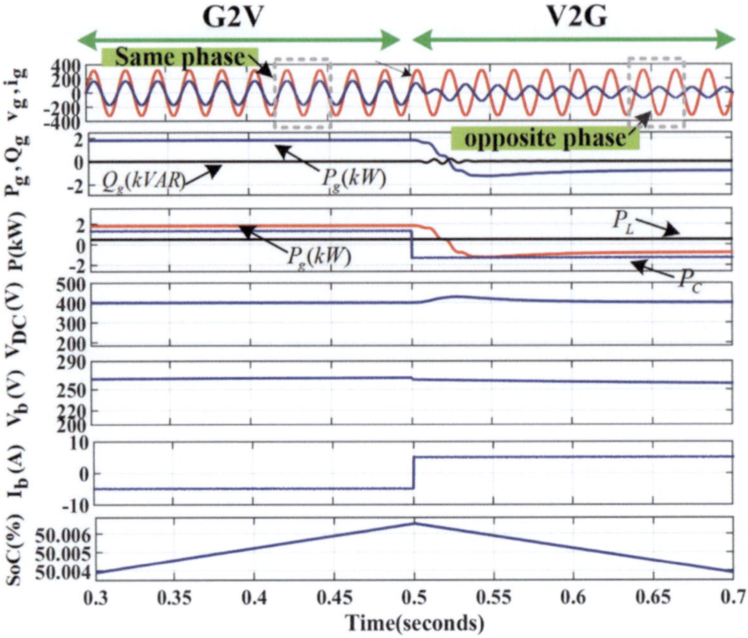

Fig. 5 Results of grid voltage (v_g), grid current (i_g), load current (i_l), and FEC current (i_{ch})

The control performance of the charger with a 10% voltage fluctuation and 15% total harmonic distortion (THD) is investigated in Figs. 6 and 7. The waveforms for grid voltage, grid current, load current, and FEC current are shown in Fig. 6. At first, the charger functions following regular grid circumstances. Between 0.15 and 0.3 s, there is a voltage decrease, and between 0.3 and 0.45 s, there is a restoration to normal voltage. There is a voltage swell between 0.45 and 0.6 s, and between 0.6 and 0.8 s, non-linear grid voltage with 15% THD is considered. The THD of grid current during G2V and V2G is 2.44% and 3.83%, respectively, as shown in Fig. 7. Figure 8 further illustrates the grid voltage, current, and active/reactive power of the FEC, load, and grid, along with the DC-link voltage, demonstrating that the grid voltage and current remain sinusoidal and at UPF. When the battery is being charged during different grid voltage disruptions, the FEC is intended to compensate for reactive power as well as load harmonics.

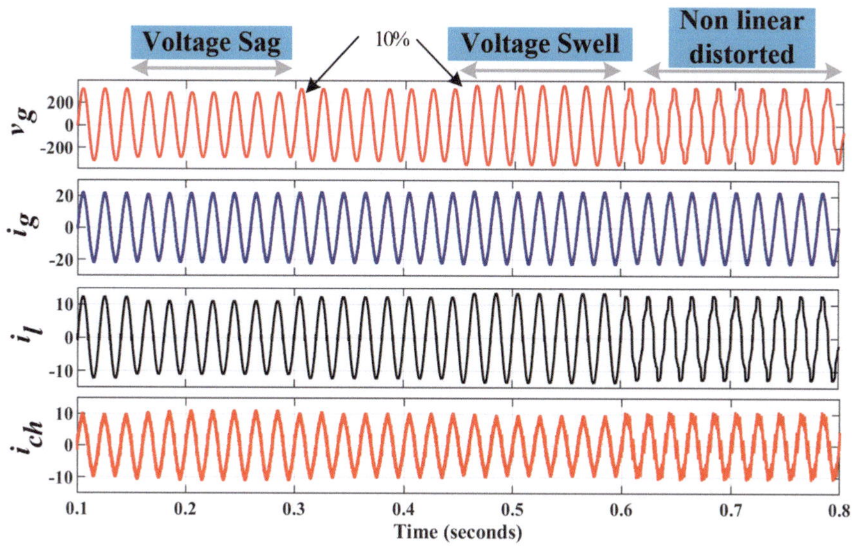

Fig. 6 Waveforms of grid voltage (v_g), grid current (i_g), load current (i_l), and FEC current (i_{ch})

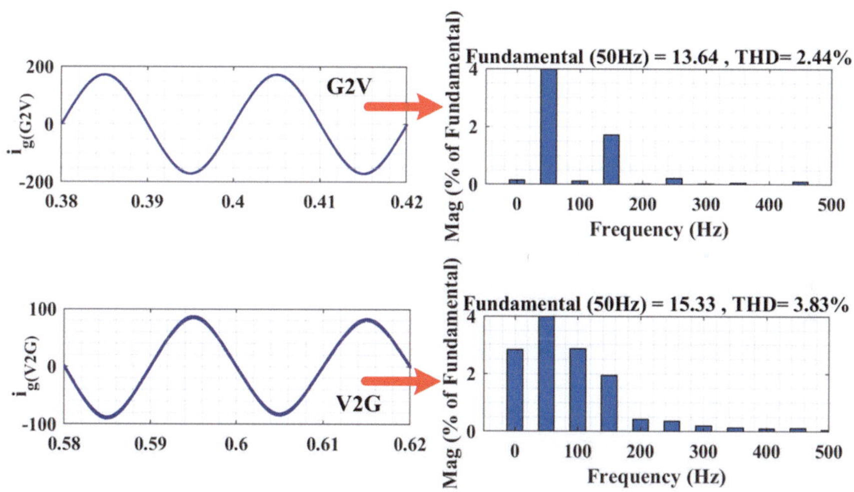

Fig. 7 Trace of grid voltage (v_g), grid current (i_g), load current (i_l), and FEC current (i_{ch})

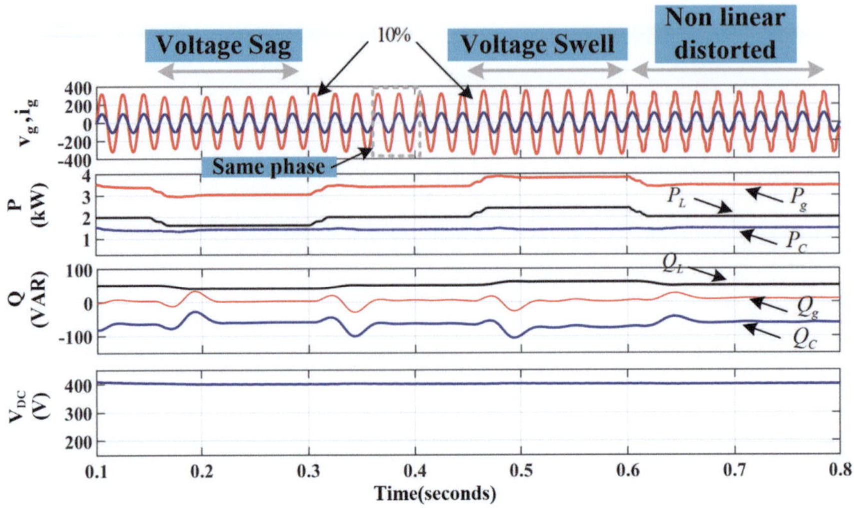

Fig. 8 Steady-state performance in distorted grid conditions

5 Conclusion

Our investigation shows that the decision tree regressor performs much better in regression tasks than both linear regression and Artificial Neural Networks (ANNs), capturing complicated data patterns and non-linear correlations. This emphasizes how crucial it is to choose models depending on the particular dataset and issue at hand rather than automatically selecting sophisticated models. The research also emphasizes how ML techniques and fourth-order quadrature signal generators (QSGs) might improve grid-interactive electric charger performance. Grid stability is improved and total harmonic distortion (THD) is decreased when traditional PI controllers are replaced with ML-based controllers, especially DT-based algorithm. These results support the integration of ML-powered technology into contemporary smart grids, enabling more dependable and effective energy management.

References

1. Ansari M, Yadav G, Singh M (2023) A 3.3kW modified LLC resonant converter for grid-tied EV system under wide voltage range. In: 2023 3rd International conference on intelligent technologies (CONIT). Hubli, India, pp 1–6. https://doi.org/10.1109/CONIT59222.2023.10205690
2. Ansari M, Yadav G, Singh M (2023) Modified bidirectional LLC converter for electric vehicle application. In: 2023 Second international conference on electrical, electronics, information and communication technologies (ICEEICT). Trichirappalli, India, pp 1–7. https://doi.org/10.1109/ICEEICT56924.2023.10157520

3. Akshay, Yadav G, Singh M (2023) Validation of a parameter design methodology for air-core two-coil WPT system with multi-resonant LC compensation. In: 2023 IEEE international conference on power electronics, smart grid, and renewable energy (PESGRE). Trivandrum, India, pp 1–6. https://doi.org/10.1109/PESGRE58662.2023.10404443
4. Yadav G, Singh M (2023) Robust control design for grid-tied EV system with SOGI-based architecture in wide voltage range scenarios. In: 2023 International conference on electrical, electronics, communication and computers (ELEXCOM). Roorkee, India, pp 1–6. https://doi.org/10.1109/ELEXCOM58812.2023.10370088
5. Kundu S, Giri AK, Kadiyan S (2025) An adaptive mixed-step size normalized least means fourth control approach for stand-alone power generation system considering dynamic conditions. IEEE J Emerg Select Topics Power Electron 13(2):2422–2429. https://doi.org/10.1109/JESTPE.2024.3516878
6. Yadav G, Singh M (2023) Unveiling the superiority: comparative analysis of ANFIS, FOPID, and PI controllers in grid-connected EV systems for G2V/V2G applications. In: 2023 International conference on electrical, electronics, communication and computers (ELEXCOM). Roorkee, India, pp 1–6. https://doi.org/10.1109/ELEXCOM58812.2023.10370196
7. Kundu S, Singh M, Giri AK et al (2024) Robust and fast control approach for islanded microgrid system and EV charging station applications. Electr Eng. https://doi.org/10.1007/s00202-024-02291-1
8. Singh A, Yadav G, Singh M (2023) Performance analysis of 3-phase active front end PWM rectifier for air-core two-coil WPT system configuration. In: 2023 9th IEEE India international conference on power electronics (IICPE). SONIPAT, India, pp 1–6. https://doi.org/10.1109/IICPE60303.2023.10474683
9. Kundu S, Giri AK, Singh M, Singh S (2025) Effective multiport battery charging facility, operation and control in hybrid isolated microgrid system. In: Giri AK, Singh M (eds) Electric vehicle charging infrastructures and its challenges. Studies in infrastructure and control. Springer, Singapore. https://doi.org/10.1007/978-981-96-0361-9_11
10. Yadav G, Singh M (2024) Real-time investigation of grid-interactive EV charger with two-stage bidirectional converter under wide voltage range scenarios. Electr Eng. https://doi.org/10.1007/s00202-024-02609-z
11. Mittal S, Singh A, Chittora P (2023) Solar PV array based grid-connected bi-directional EV charger controlled using NARLMMN algorithm. In: 2023 9th IEEE India international conference on power electronics (IICPE). IEEE
12. Kundu S, Giri AK, Singh M et al (2024) Adaptive control strategy for isolated renewable energy-based generation system with intermittency. Electr Eng. https://doi.org/10.1007/s00202-024-02454-0
13. Siddiqui MK, Yadav G, Wasti AS, Ansari MA (2022) Performance analysis of power flow controller used for solar PV based grid integrated system. In: 2022 2nd International conference on emerging frontiers in electrical and electronic technologies (ICEFEET). Patna, India, pp 1–5. https://doi.org/10.1109/ICEFEET51821.2022.9847906
14. Singh S et al (2024) Design and control of electric vehicle charger using bi-directional DC-DC CUK converter. In: 2024 IEEE Third international conference on power electronics, intelligent control and energy systems (ICPEICES). IEEE
15. Kundu S, Singh M, Giri AK (2023) Adaptive control approach-based isolated microgrid system with alleviating power quality problems. Electric Power Components Syst 52(7):1219–1234. https://doi.org/10.1080/15325008.2023.2239222
16. Kundu S, Singh M, Giri AK (2022) Control algorithm for coordinated operation of wind-solar microgrid standalone generation system. Energy Sour Part A Recover Utili Environ Effects 44(4):10024–10044. https://doi.org/10.1080/15567036.2022.2143943
17. Mittal S, Singh A, Chittora P (2024) Design and development of Reduced Switch Five Level Inverter for interfacing renewable energy sources and EV charger. Electr Eng 1–15
18. Mittal S, Singh A, Chittora P (2022) EV control in G2V and V2G modes using SOGI controller. In: 2022 IEEE 3rd global conference for advancement in technology (GCAT). IEEE

19. Mittal S, Singh A, Chittora P (2024) Power quality enhancement in single phase two level/five level converters using adaptive-RBFNN algorithm. Electr Eng 1–14
20. Mittal S, Singh A, Chittora P (2024) Design and development of leaky least mean fourth control algorithm for single-phase grid-connected multilevel inverter. Int J Circuit Theory Appl 52(1):328–345
21. Mohan KB, Sreedhar M, Shanthi K, Nandavaram B (2023) 3. Implementation of deep learning-based Bi-directional DC-DC converter for V2V and V2G applications—an experimental investigation. Energies. https://doi.org/10.3390/en16227614
22. Nilgun F-U, Melike YA (2023) 5. Electric vehicle charging service operations: a review of machine learning applications for infrastructure planning, control, pricing and routing. Renewable & sustainable energy reviews. https://doi.org/10.1016/j.rser.2023.113873
23. Syed HA, Iftikhar A (2024) 1. Optimal wireless power transfer to hybrid energy storage system for electric vehicles: a comparative analysis of machine learning-based model-free controllers. J Energy Storag. https://doi.org/10.1016/j.est.2023.109534
24. Kundu S, Singh M, Giri AK (2024) Synchronization and control of WECS-SPV-BSS-based distributed generation system using ICCF-PLL control approach. Electr Power Syst Res 226:109919. ISSN 0378-7796. https://doi.org/10.1016/j.epsr.2023.109919
25. Kundu S, Giri AK, Kadiyan S, Singh S (2025) Power management and adaptive control method for a standalone distributed generation system under fault, DC-offset, and dynamic environments. Int J Ambient Energy 46(1). https://doi.org/10.1080/01430750.2025.2452245
26. Singh N, Pratap B (2025) Robust observer design for variable speed wind turbine using a two-mass model. In: Proceedings of the international conference on systems, control and automation. ICSCA 2023. Lecture notes in electrical engineering, vol 1260. Springer, Singapore. https://doi.org/10.1007/978-981-97-7384-8_15

Chapter 5
Integration of Fuel Cell–Renewable Energy Sources in Hybrid Infrastructure

Preeti Sharma, Shubham Priyadarshi, Anakshi Boruah, and Rajneesh Kumar

Abstract The integration of fuel cells with renewable energy sources offers a sustainable and efficient solution for power generation. By combining hydrogen-based fuel cells with solar and wind energy, the system ensures enhanced reliability and energy storage capabilities. This hybrid approach provides a consistent power supply during low-generation periods. It also helps reduce greenhouse gas emissions and dependence on fossil fuels. Advanced control strategies play a crucial role in optimizing system performance and managing the variability of renewable sources. Additionally, fuel cells offer fast response times and high efficiency, making them ideal for hybrid configurations. The stored hydrogen can be utilized when renewable generation is low, ensuring energy availability. This setup also supports grid stability and can be scaled for both off-grid and grid-connected applications. Integrating smart energy management systems further improves the coordination among sources. Overall, this hybrid infrastructure supports the transition toward cleaner and more resilient energy systems.

Keywords Fuel cell · Renewable energy · Hybrid systems · Hydrogen storage · Energy management · Solar PV · Wind energy · Electrolyzer · Smart grid · Sustainable energy

P. Sharma (✉)
Department of Electrical Engineering, Pandit Deendayal Energy University, Gujarat, Gandhinagar, India
e-mail: preetisharma3782@gmail.com

S. Priyadarshi · R. Kumar
Department of Electrical and Electronics Engineering, Birla Institute of Technology and Science BITS Pilani, Pilani, Rajasthan, India
e-mail: p20230039@pilani.bits-pilani.ac.in

R. Kumar
e-mail: rajneesh@pilani.bits-pilani.ac.in

A. Boruah
Department of Chemistry, Birla Institute of Technology and Science BITS Pilani, Pilani, Rajasthan, India
e-mail: p20230014@pilani.bits-pilani.ac.in

S. Kundu et al. (eds.), *Hybrid Renewable Power Infrastructure for Sustainable Electric Vehicle Development*, Studies in Infrastructure and Control,
https://doi.org/10.1007/978-981-95-1323-9_5

1 Introduction

The growing global energy demand, coupled with the urgency to reduce greenhouse gas emissions, has driven a paradigm shift toward renewable energy sources. However, the inherent intermittency of solar and wind power creates challenges in maintaining a consistent and reliable power supply [1]. Integrating fuel cells with renewable energy sources within hybrid infrastructures offers a promising and sustainable solution to these issues. Fuel cells, which efficiently convert hydrogen into electricity with water as the only byproduct, can serve as a clean backup or complementary source to fluctuating renewables. This integrated approach enhances system resilience, supports energy storage, and ensures continuous energy availability [2, 3].

The concept of hybrid infrastructure also enables a dynamic energy ecosystem where green hydrogen can be produced using excess energy from renewable sources through electrolysis. This stored hydrogen becomes a crucial asset, enabling load balancing and peak shaving during periods of high demand or low renewable generation. Additionally, the integration promotes distributed generation, reduces reliance on centralized fossil-based power systems, and provides energy access to remote or disaster-prone regions [4, 5].

As technological innovations advance, hybrid systems incorporating intelligent energy management, advanced control algorithms, and real-time monitoring are becoming more feasible and economically viable. These systems not only mitigate the limitations of standalone renewable technologies but also contribute to global decarbonization goals. This chapter delves into the components, architecture, benefits, control mechanisms, and future directions of fuel cell–renewable hybrid systems, supported by recent advances in research and technological innovation.

Refer to Fig. 1, which presents a schematic overview of the energy interaction and integration between various components in a fuel cell–renewable hybrid system. It highlights the roles of solar PV, wind turbines, electrolyzers, hydrogen storage, batteries, fuel cells, and inverters under the coordination of an energy management system (EMS) [6, 7].

This chapter contains an overview of fuel cells in Sect. 2, followed by a discussion on the role of renewable energy sources in hybrid systems in Sect. 3. Section 4 presents the overall architecture and configuration of hybrid energy systems. Section 5 explores energy management strategies and control techniques. Section 6 discusses recent research trends, case studies, and challenges related to hybrid system deployment. The chapter concludes with Sect. 7, offering a forward-looking perspective on the scalability, policy implications, and economic feasibility of such hybrid infrastructures in shaping a sustainable energy future.

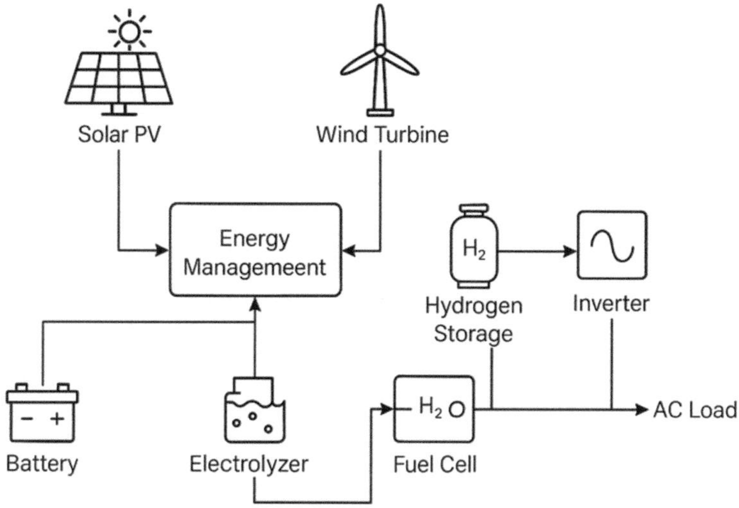

Fig. 1 Schematic diagram of a fuel cell–renewable hybrid energy system

2 Fuel Cells: A Brief Overview

Fuel cells are electrochemical energy conversion devices that generate electricity by facilitating a reaction between hydrogen and oxygen. Unlike traditional combustion-based power systems, fuel cells do not rely on burning fuel, which significantly reduces harmful emissions and enhances overall efficiency [2, 8]. The only byproducts in hydrogen-fueled fuel cells are water and heat, making them an environmentally friendly option.

There are several types of fuel cells, each suitable for different applications based on their operating temperatures, electrolyte materials, and efficiency characteristics:

- **Proton Exchange Membrane Fuel Cells (PEMFCs)**—Operate at relatively low temperatures (~60–80 °C), ideal for residential and automotive applications due to their quick startup time and high power density.
- **Solid Oxide Fuel Cells (SOFCs)**—Function at high temperatures (700–1000 °C), suitable for stationary power generation, especially in combined heat and power (CHP) systems.
- **Alkaline Fuel Cells (AFCs)** and **Phosphoric Acid Fuel Cells (PAFCs)**—Traditionally used in space missions and commercial buildings, offering stable performance over extended durations.

In integrated systems, fuel cells offer continuous and stable power generation, especially during periods of reduced solar irradiance or low wind speeds. When linked with renewable sources, fuel cells can act as both a backup and a primary generator depending on energy availability and demand.

A critical component of these hybrid setups is the electrolyzer, which converts surplus renewable electricity into hydrogen. This hydrogen is stored in tanks and supplied to fuel cells when necessary, establishing a sustainable cycle. The modular nature of fuel cells allows for scalability, adapting to small residential loads or large industrial demands with ease.

The benefits of fuel cells extend beyond clean electricity. Their silent operation, minimal maintenance, and resilience to environmental conditions make them ideal for deployment in hospitals, data centers, remote areas, and disaster recovery zones. Additionally, fuel cell systems can be co-located with renewable plants or integrated into microgrids, enhancing energy autonomy and local energy security [9–11].

Figure 2 illustrates the working principle of a hydrogen fuel cell. Hydrogen gas enters the anode, where it is split into protons and electrons. The electrons pass through an external circuit to generate electricity, while protons move through an electrolyte to the cathode. At the cathode, they recombine with oxygen and electrons to form water [12, 13].

DC bus is not connected to electrolyzer directly because of its inability to control the power flow. Power conversion stages from source to load require power electronics application. AC-to-DC converter and DC-to-DC converter are required in case the electrolyze is supplied from grid or solar photovoltaic, respectively, as shown in figure below. An electrolyzer is a high-current low-voltage DC system in which current determines the rate of hydrogen production and voltage drives the electrolysis process. DC–DC converter should have voltage step-down ratio of 8:1 or 10:1 when interfacing it between source (rooftop solar or DC microgrid) and electrolyzer as shown in Fig. 3a and b. The requirements of DC–DC converter for renewable energy integration with electrolyzer are as follows:

Fig. 2 Working principle of a hydrogen fuel cell

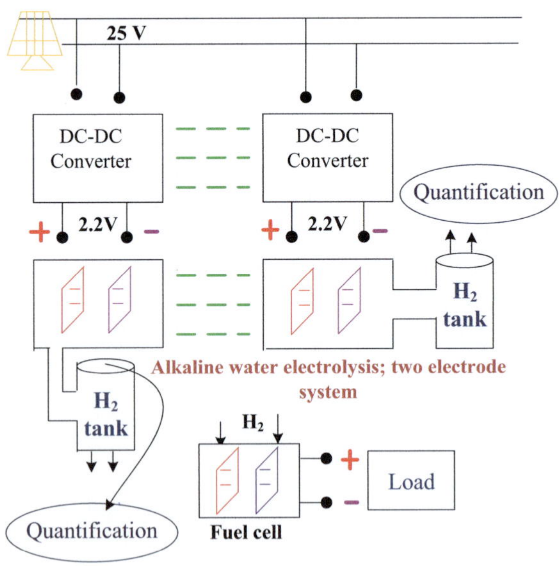

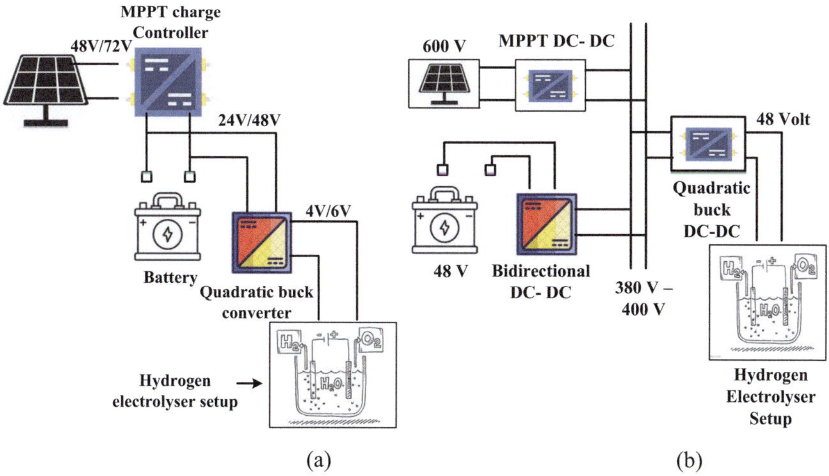

Fig. 3 **a** Green hydrogen production with low power system (using solar and battery), **b** Green hydrogen production with high power system (using DC microgrid)

(a) Large voltage step-down capability.
(b) Low output current ripple.
(c) High efficiency.
(d) Low electromagnetic interference.

Electrochemical water splitting has potential for sustainable H_2 production, offering benefits such as zero carbon emissions and high energy density. Conventional hydrogen production methods, like steam reforming, release CO_2 and are not sustainable. The two half-cell reactions active in the electrochemical water splitting process are HER and OER. In this process, H_2O is dissociated into hydrogen (H_2) and oxygen (O_2) gases when an external electrical voltage, which is applied across the system. The efficiency is largely confirmed by the electrocatalyst's ability to facilitate these reactions [14]. The hydrogen generated by electrolysis of water is termed as green if the input energy required is supplied from renewable sources like solar power or tidal, having zero carbon emissions. Photovoltaic-electrochemical (PV-EC) water splitting produces green hydrogen as it integrates solar energy with electrochemical cells. In PV-EC water splitting, the input solar energy provides the voltage that drives the electrolysis process and the current generated determines the rate of hydrogen production. PV-EC systems are ecofriendly, offer low fuel costs, and are durable, making them a promising renewable energy solution [8, 15, 16].

In electrochemical measurements, two-electrode and three-electrode setups are frequently utilized as shown in Fig. 4. Three electrodes make up a three-electrode system: a working electrode, which is where the reaction of interest takes place; a reference electrode, which offers a steady potential for precise measurement; and a counter-electrode, also known as an auxiliary electrode, which completes the circuit

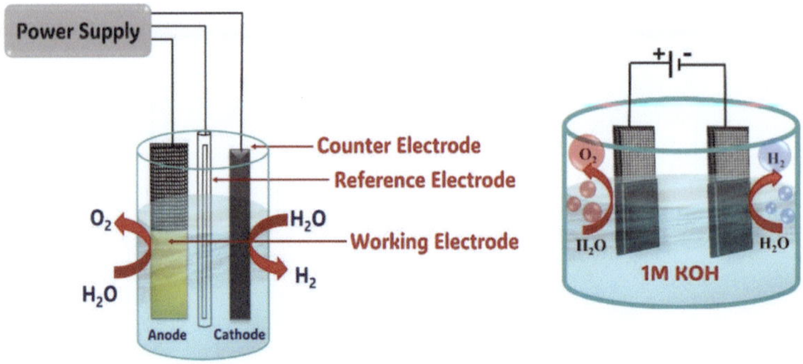

Fig. 4 Three-electrode system and two-electrode system

and conducts current. This configuration is perfect for methods like chronoamperometry and cyclic voltammetry because it enables accurate control and measurement of the working electrode potential. A two-electrode system, on the other hand, only uses a working electrode and a combined counter/reference electrode. Applications like batteries and electroplating, where exact potential control is not essential, usually employ this less complicated configuration. But it isn't as accurate as a three-electrode system because current flow can cause the reference electrode potential to change [17, 18].

3 Renewable Energy Sources in Hybrid Systems

Renewable energy sources, particularly solar and wind, are central to hybrid system configurations. Their role in creating sustainable, low-emission energy systems is critical; however, their intermittency requires the incorporation of auxiliary systems like fuel cells to maintain a steady power supply. Solar and wind energies have distinct generation profiles and complementary behaviors, making them suitable for hybridization [19].

- **Solar Photovoltaic (PV) Systems** convert sunlight directly into electricity using semiconductor materials. These systems are easy to install, scalable, and maintenance-light. However, they depend heavily on weather conditions and diurnal cycles, resulting in fluctuating outputs.
- **Wind Turbines** harness the kinetic energy of wind to generate electricity. Wind energy production can occur at night or during cloudy days when solar generation is low, thus complementing solar PV. However, wind patterns are site-specific and subject to seasonal changes.

Individually, these systems are intermittent and cannot guarantee continuous power. However, when used together, solar and wind can reduce the total variability

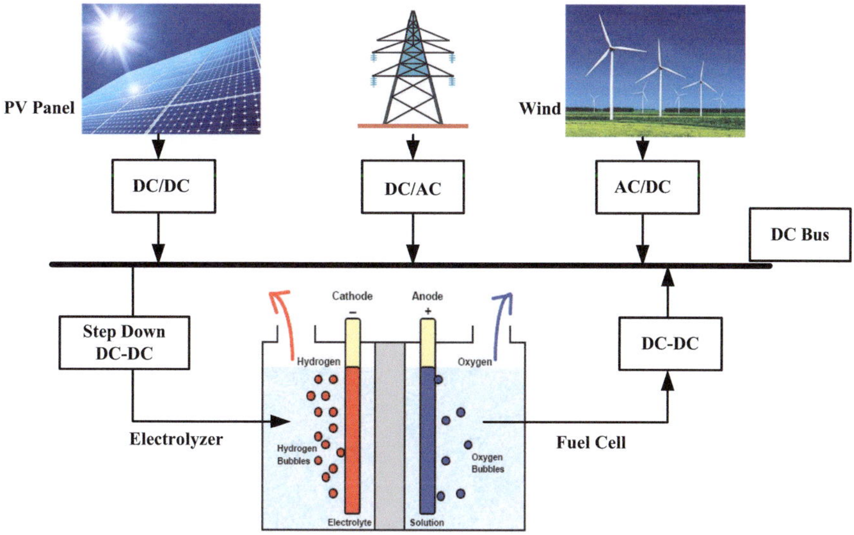

Fig. 5 Block diagram of a hybrid renewable energy system with fuel cell integration

of renewable output. Coupling them with storage systems or dispatchable power sources such as fuel cells enhances the overall system reliability and flexibility.

Advanced hybrid systems incorporate battery banks, electrolyzers, and hydrogen storage units. During periods of excess renewable generation, electricity is stored either in batteries or used to produce hydrogen. This hydrogen can later be converted back into electricity using fuel cells. This cycle not only stores energy efficiently but also decouples generation from consumption time.

Hybrid renewable systems with fuel cells are deployed in various contexts, from off-grid rural electrification to large-scale urban microgrids. Their flexibility and modularity allow for customized configurations based on local resource availability, energy demand patterns, and environmental considerations.

Figure 5 illustrates a hybrid energy system combining solar, wind, and fuel cells, including power conditioning units and storage components. It highlights the interplay between renewable sources, storage, hydrogen production, and fuel cell-based power generation [5, 6, 20].

4 Overall Architecture of Hybrid Energy Systems

The architecture of hybrid energy systems integrating fuel cells and renewables is based on a multi-layered design that supports power generation, energy conversion, and intelligent control as shown in Fig. 6. The typical components include renewable energy generators (solar PV, wind turbines), electrolyzers, hydrogen storage, fuel

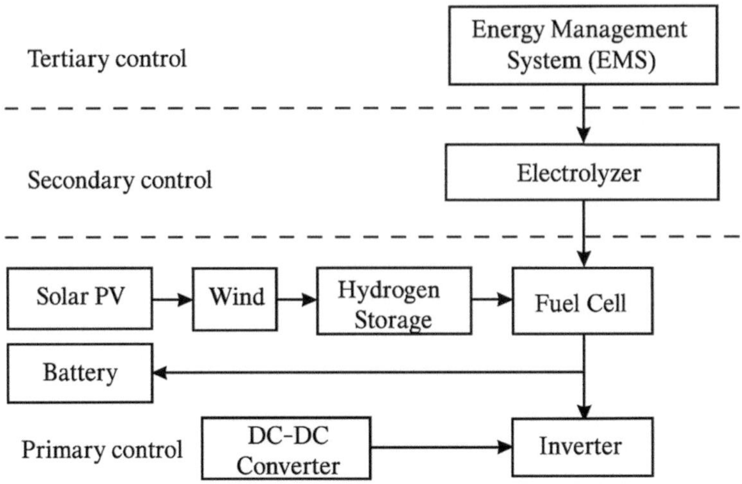

Fig. 6 Architecture of fuel cell–renewable hybrid energy systems

cells, battery banks, DC–DC converters, inverters, and an energy management system (EMS).

These systems can be configured in either AC-coupled or DC-coupled architectures. DC-coupled systems offer higher efficiency by reducing the number of power conversions, making them suitable for systems with extensive battery storage. In contrast, AC-coupled systems provide better flexibility for integrating conventional loads and grid connections.

A hierarchical control architecture is often implemented. At the primary level, converters manage local voltage and current. The secondary level involves coordination among devices to maintain system stability and optimize energy flow [21]. The tertiary level includes the EMS, which forecasts demand, schedules generation, and decides when to store or dispatch energy.

Redundant paths and modular design elements enhance reliability and scalability. Microgrids based on this architecture can operate in both grid-connected and islanded modes, supporting resilience during grid failures or peak load conditions. Smart meters and IoT-based sensors ensure real-time monitoring and adaptive control [22, 23].

5 Energy Management Strategies and Control Techniques

Efficient energy management is essential for the reliable operation of hybrid systems. The EMS balances generation, consumption, and storage using real-time data and predictive analytics. It aims to minimize operational costs, maximize renewable utilization, and maintain power quality.

Control techniques used in these systems can be classified as rule-based, optimization-based, or intelligent methods. Rule-based strategies rely on predefined logic to control energy flow. While simple to implement, they may not be optimal under varying conditions. Optimization-based techniques, such as linear programming and model predictive control (MPC), provide dynamic control by solving objective functions subject to system constraints [24, 25].

Artificial intelligence (AI) and machine learning (ML) are increasingly applied to forecast renewable generation, predict load demand, and adapt control policies. Fuzzy logic controllers and neural networks enable robust decision-making under uncertainty. Hybrid controllers combining rule-based and intelligent methods are also emerging as promising solutions.

Communication between EMS components is facilitated by standard protocols such as Modbus, IEC 61,850, and MQTT. Cybersecurity measures are vital to ensure system integrity, particularly in smart grid and IoT-enabled environments [26].

6 Case Studies, Challenges, and Research Trends

Numerous pilot projects and real-world deployments have validated the performance of fuel cell–renewable hybrid systems. Examples include remote microgrids in Alaska, hydrogen-powered smart homes in Japan, and integrated urban microgrids in Germany and the Netherlands.

Key challenges include high initial capital costs, limited hydrogen infrastructure, efficiency losses during energy conversion, and system complexity. The development of durable and cost-effective electrolyzers and fuel cell stacks remains a focus of research as shown in Fig. 7. Additionally, improving the lifecycle sustainability of components and recycling strategies is critical.

Recent trends include the integration of electric vehicles (EVs) as mobile storage units, peer-to-peer energy trading, and blockchain-based energy transactions. Research is also progressing toward solid-state hydrogen storage, hybrid AC/DC microgrids, and seamless multi-energy coupling [2, 27–30].

7 Conclusion and Future Outlook

The integration of fuel cells with renewable energy systems marks a significant advancement in the development of clean, dependable, and resilient energy infrastructures. By mitigating the variability associated with intermittent renewable sources and enabling long-duration energy storage through hydrogen, these hybrid systems offer a flexible and efficient energy solution. Their applicability spans a broad spectrum—from remote and off-grid locations to complex, urbanized power networks.

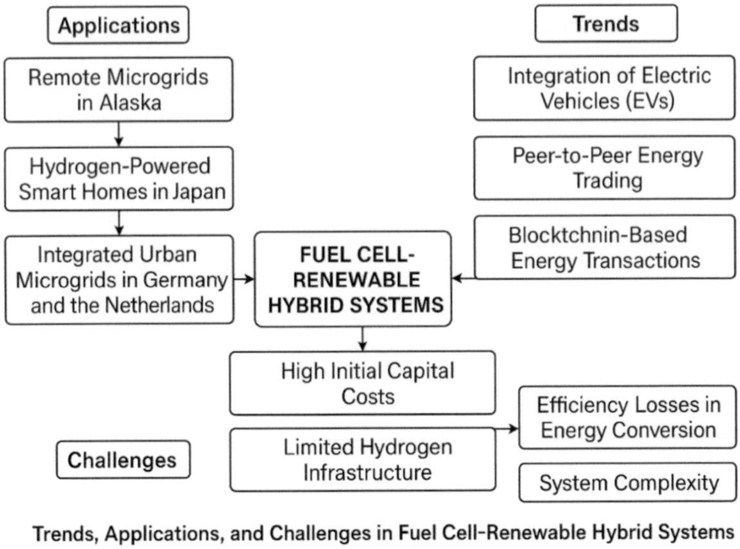

Fig. 7 Trends, applications, and challenges in fuel cell–renewable hybrid systems

The combination of electrochemical conversion via fuel cells with variable renewable energy enhances the overall system efficiency and operational reliability while supporting decarbonization efforts.

Despite their potential, several techno-economic barriers must be addressed to facilitate large-scale deployment. These include the high capital cost of fuel cell systems, limited infrastructure for hydrogen production and storage, and challenges associated with dynamic system integration and control. Future research should focus on improving the cost-effectiveness and performance metrics of proton exchange membrane (PEM) and solid oxide fuel cells (SOFCs), enhancing the efficiency of electrolysis processes for green hydrogen production, and developing robust power management algorithms to ensure real-time system adaptability. Policy interventions and financial incentives will also play a critical role in accelerating the adoption of such hybrid configurations.

Advancements in artificial intelligence (AI)-driven energy management systems and predictive control techniques can substantially enhance the performance of hybrid systems. By incorporating forecasting models for renewable energy generation and load profiles, these systems can achieve optimal energy dispatch, fault tolerance, and lifecycle cost reduction. Furthermore, the integration of hybrid energy systems with electric vehicle charging infrastructure, smart grids, and decentralized energy markets can offer additional benefits such as peak load shaving, grid stability, and enhanced user participation in energy trading.

In conclusion, hybrid infrastructures incorporating fuel cells and renewable energy sources will be integral to the realization of carbon-neutral energy systems. As global energy paradigms shift toward decentralization, decarbonization, and digitalization, such systems will form the backbone of future sustainable energy networks. Continued collaboration among researchers, policymakers, and industry stakeholders is imperative to overcome existing limitations and unlock the full potential of hybrid fuel cell–renewable systems in achieving long-term climate and energy security goals.

References

1. A Pramuanjaroenkij S Kakaç 2023 The fuel cell electric vehicles: the highlight review Int J Hydrogen Energy 48 25 9401 9425 https://doi.org/10.1016/j.ijhydene.2022.11.103
2. N Sazali WNW Salleh AS Jamaludin MNM Razali 2020 New perspectives on fuel cell technology Membranes (Basel) 10 5 99
3. RW Erickson D Maksimovic 2001 Fundamentals of power electronics Kluwer Academic Publishers New York
4. P Garcia LM Fernandez CA Garcia F Jurado 2010 Energy management system of fuel-cell-battery hybrid tramway IEEE Trans Ind Electron 57 12 4013 4023 https://doi.org/10.1109/TIE.2009.2034173
5. Amusan OT, Nwulu NI, Gbadamosi SL (2024) Optimal solar-biomass-diesel-generator hybrid energy for water pumping system considering demand response. Eng Reports 1–14. https://doi.org/10.1002/eng2.13055
6. S Kundu M Singh AK Giri 2021 Design and control of a standalone wind-solar system with SEIG feeding linear/nonlinear loads Lect Notes Electr Eng. 616 317 327 https://doi.org/10.1007/978-981-16-1978-6_28
7. S Kundu M Singh AK Giri 2024 SPV-wind-BES-based islanded electrical supply system for remote applications with power quality enhancement Electr Eng 106 1 279 294 https://doi.org/10.1007/s00202-023-01979-0
8. MK Singla P Nijhawan AS Oberoi 2021 Hydrogen fuel and fuel cell technology for cleaner future: a review Environ Sci Pollut Res 28 13 15607 15626 https://doi.org/10.1007/s11356-020-12231-8
9. S Kundu M Singh AK Giri S Kadiyan P Chittora D Shailly 2024 Robust and fast control approach for islanded microgrid system and EV charging station applications Electr Eng https://doi.org/10.1007/s00202-024-02291-1
10. S Kundu AK Giri M Singh S Kadiyan 2024 Adaptive control strategy for isolated renewable energy-based generation system with intermittency Electr Eng https://doi.org/10.1007/s00202-024-02454-0
11. Sharma P, Hasanpour S, Kumar R A new soft-switching high step-up DC-DC converter with low voltage stress statements and declarations : Funding : The authors did not receive support from any organization for the submitted work. Financial interests : The authors declare they have no, pp 1–12
12. S Mekhilef R Saidur A Safari 2012 Comparative study of different fuel cell technologies Renew Sustain Energy Rev 16 1 981 989 https://doi.org/10.1016/j.rser.2011.09.020
13. Bizon N, Stan VA, Cormos AC (2019) Optimization of the fuel cell renewable hybrid power system using the control mode of the required load power on the DC bus. Energies 12(10). https://doi.org/10.3390/en12101889
14. Rezk H, Fathy A (2024) Hydrogen reduction-based energy management strategy of hybrid fuel cell/PV/battery/supercapacitor renewable energy system. J Energy Storage 86:111316. https://doi.org/10.1016/j.est.2024.111316
15. O Atlam M Kolhe 2011 Equivalent electrical model for a proton exchange membrane (PEM) electrolyser Energy Convers Manag 52 8–9 2952 2957 https://doi.org/10.1016/j.enconman.2011.04.007

16. SA Gorji 2022 Reconfigurable quadratic converters for electrolyzers utilized in DC Microgrids IEEE Access 10 October 109677 109687 https://doi.org/10.1109/ACCESS.2022.3214581
17. D Yao L Gu B Zuo S Weng S Deng W Hao 2021 A strategy for preparing high-efficiency and economical catalytic electrodes toward overall water splitting Nanoscale 13 24 10624 10648 https://doi.org/10.1039/d1nr02307a
18. Z Wu 2025 A comprehensive review of modeling of solid oxide fuel cells: from large systems to fine electrodes Chem Rev https://doi.org/10.1021/acs.chemrev.4c00614
19. Obileke KC, Mukumba P (2025) Techno-economic evaluation of wind and bio-energy systems for sustainable development: a systematic review. Energy Sci Eng 2179–2202. https://doi.org/10.1002/ese3.70025
20. Mumtaz F, Zaihar Yahaya N, Tanzim Meraj S, Singh B, Kannan R, Ibrahim O (2021) Review on non-isolated DC-DC converters and their control techniques for renewable energy applications. Ain Shams Eng J 12(4):3747–3763. https://doi.org/10.1016/j.asej.2021.03.022
21. Ghoniem RM, Alahmer A, Rezk H, As'ad S (2023) Optimal design and sizing of hybrid photovoltaic/fuel cell electrical power system. Sustain 15(15). https://doi.org/10.3390/su151512026
22. Ghenai C, Bettayeb M, Brdjanin B, Hamid AK (2019) Hybrid solar PV/PEM fuel Cell/Diesel Generator power system for cruise ship: a case study in Stockholm, Sweden. Case Stud Therm Eng 14:100497. https://doi.org/10.1016/j.csite.2019.100497
23. M Trifkovic M Sheikhzadeh K Nigim P Daoutidis 2014 Modeling and control of a renewable hybrid energy system with hydrogen storage IEEE Trans Control Syst Technol 22 1 169 179 https://doi.org/10.1109/TCST.2013.2248156
24. TS Babu KR Vasudevan VK Ramachandaramurthy SB Sani S Chemud RM Lajim 2020 A comprehensive review of hybrid energy storage systems: converter topologies, control strategies and future prospects IEEE Access 8 148702 148721 https://doi.org/10.1109/ACCESS.2020.3015919
25. Bizon N, Thounthong P (2020) Energy efficiency and fuel economy of a fuel cell/renewable energy sources hybrid power system with the load-following control of the fueling regulators. Mathematics 8(2). https://doi.org/10.3390/math8020151
26. Y Zhang C Zhang Z Huang L Xu Z Liu M Liu 2019 Real-time energy management strategy for fuel cell range extender vehicles based on nonlinear control IEEE Trans. Transp. Electrif. 5 4 1294 1305 https://doi.org/10.1109/TTE.2019.2958038
27. J Ebersberger M Hagedorn M Lorenz A Mertens 2022 Potentials and comparison of inverter topologies for future all-electric aircraft propulsion IEEE J Emerg Sel Top Power Electron 10 5 5264 5279 https://doi.org/10.1109/JESTPE.2022.3164804
28. Singh N, Pratap B (2025) Robust observer design for variable speed wind turbine using a two-mass model. In: Proceedings of the international conference on systems, control and automation. ICSCA 2023. Lecture notes in electrical engineering, vol 1260. Springer, Singapore. https://doi.org/10.1007/978-981-97-7384-8_15
29. S Kundu AK Giri S Kadiyan 2025 An adaptive mixed-step size normalized least means fourth control approach for stand-alone power generation system considering dynamic conditions IEEE J Emerg Sel Top Power Electron 13 2 2422 2429 https://doi.org/10.1109/JESTPE.2024.3516878
30. Kundu S, Giri AK, Kadiyan S, Singh S (2025) Power management and adaptive control method for a standalone distributed generation system under fault, DC-offset, and dynamic environments. Int J Ambient Energy 46(1). https://doi.org/10.1080/01430750.2025.2452245

Chapter 6
Power Quality Enhancement in Grid-Connected EV Charging Systems Using Third-Order Sinusoidal Integrators to Control BADC

Lochan Varshney, Chaudhary Indra Kumar, and Prakash Chittora

Abstract New energy exchange systems between EVs and the power grid are made possible by the evolution of electric vehicle technology. Two such developments are the bidirectional power flow made possible by Vehicle-to-Grid (V2G) and Grid-to-Vehicle (G2V) systems. EVs may replenish stored energy for the grid in V2G mode; in G2V mode, they utilize grid power to charge their batteries. This capacity helps with load balancing, voltage control, and general system dependability, therefore supporting grid stability. This work proposes a dual-mode onboard battery charger suitable for V2G and G2V operations. The suggested approach connects EV to the grid via bidirectional converters—more specifically, a bidirectional AC–DC/DC–DC converter. Third-Order Sinusoidal Signal Integrator (TOSSI) control approach manages the bidirectional AC-to-DC converter, therefore guaranteeing good steady-state and dynamic performance. Furthermore, under control while charging and discharging are the voltage and current of the battery via a PWM-based controller. Additionally examined in this work is how fluctuations in nonlinear load impact system behavior. Extensive simulations run in the MATLAB/Simulink environment confirm the proposed method.

Keywords TOSSI controller · Bidirectional converter · *G2V-V2G* · Second-order generalized integrator (SOGI)

L. Varshney (✉) · C. I. Kumar · P. Chittora
Electrical Engineering, Delhi Technological University, Delhi, India
e-mail: varshneylochan@gmail.com

C. I. Kumar
e-mail: indrachaudhry@dtu.ac.in

P. Chittora
e-mail: prakashchittora@dtu.ac.in

S. Kundu et al. (eds.), *Hybrid Renewable Power Infrastructure for Sustainable Electric Vehicle Development*, Studies in Infrastructure and Control,
https://doi.org/10.1007/978-981-95-1323-9_6

1 Introduction

The world's population explosion and the depletion of conventional fossil fuel supply have made environmentally friendly and energy-efficient transportation choices more important. Reduced dependence on fossil fuels and lower greenhouse gas emissions have made electric vehicles (EVs) a reasonable replacement for cars using internal combustion engines. With a 90% compound annual growth rate, emerging countries like India, which intend to spend $150 billion in battery-integrated electric vehicle development by 2030, show clear changes [1, 2].

An essential component allowing EVs to be connected to the electric grid is the onboard battery charger (OBC), which lets vehicles draw electricity directly from the AC grid. OBCs are preferred in the automobile sector over off-board charging devices, which are often bigger and more expensive, because of their mobility and ease of use [3, 4].

Unidirectional OBCs are usually used because of their simple design and little influence on battery life; yet, they are not suitable for more complicated characteristics like V2G operations [5]. Research on bidirectional OBCs, which let EVs send power back into the grid during times of peak demand, has taken the front stage to facilitate bidirectional energy flow, which is essential for V2G systems [6, 7]. Usually using bidirectional switches, i.e., insulated transistors (IGBTs), these systems may operate in both G2V and V2G modes.

The enormous amount of energy that EV batteries can store will open up possibilities for dynamic energy management and grid support as the EV market expands, particularly during emergencies or power outages. Control strategies and intelligent power electronics become critical in this situation. The different novel control approaches, like dual third-order generalized integrator-based phase-locked loop (DTOGI-PLL) [8], modified proportionate affine projection algorithm (MPAPA) [9], mixed-step size normalized least means fourth (MSSNLMF) [10], leaky least mean fourth (LLMF) [11], adaptive radial basis functional neural network (ARBFNN) [12], second-order sequence filter-based frequency locked loop (SOSF-FLL) [13], and brain-inspired learning (BIL) [14], are used to control power electronics-based voltage source converters in grid- or generator-connected EV systems.

Much research has proposed improvements in bidirectional charging systems. For example, whereas a programmable DC link voltage has been utilized to lower high-frequency ripple in the grid [15], the SOGI approach has been used to increase system stability and dependability [16]. Other approaches, including novel converters using TOSSI control methods and expanding functionality to include charge and discharge systems [4], highlight even more the versatility of modern EV chargers. The light electric vehicles using permanent magnet synchronous motors are also designed to mitigate thermal effects [17, 18], and an efficient motor with advanced converters is also designed for smooth vehicle drive [19–21].

Concerning converter design, it investigates a two-converter system [22] sharing a DC connection. One converter in this study is connected to the grid and the other to the battery. Whereas the bidirectional DC–DC converter (BDDC) controls battery charging, the grid-side converter (GSC), implemented as a bidirectional AC–DC converter, guarantees a unity power factor (UPF). Furthermore, for effective energy flow across many load conditions, research for effective energy transfer under various load scenarios has been focused on other improvements, such as Power Factor Correction (PFC) and Inductor-Inductor-Capacitor (LLC) topologies. Despite their advanced features, bidirectional OBCs still have many drawbacks, such as poor power density, higher weight, and high cost [23, 24]. Nonetheless, the implementation of V2G technology holds significant promise for enhancing grid dependability and flexibility [25].

In addition to innovations in charger design targeted at lowering size, cost, and energy loss, studies like [26, 27] have highlighted industry-standard charging power levels, different charger topologies, and difficulties in EV/PHEV infrastructure.

The current study expands on this body of work by suggesting an onboard bidirectional battery charger for EVs that facilitates both V2G (vehicle-to-grid) and G2V (grid-to-vehicle) operations. The fundamental current needed to produce reference signals for the BADC, which doubles as a shunt compensator, is extracted by the system using the Third-Order Sinusoidal Signal Integrator (TOSSI) technique. By lowering harmonics and raising the power factor, this control approach improves power quality and permits smooth bidirectional energy transfer between the grid and the car.

2 Proposed System Configuration

The general layout of the suggested electric vehicle charging system is depicted in Fig. 1. From Fig. 1, it is clear that the system is largely composed of two power electronics converters: a BADC and BDDC. A shared DC link, which serves as an energy transfer bridge between the grid and the vehicle battery, connects these two converters.

In Grid-to-Motor (G2M) operation, the BADC functions as a rectifier, converting AC power from the grid into DC, which then charges the battery through the DC link capacitor. This capacitor not only smooths the DC voltage but also filters out high-frequency harmonics introduced by switching operations. Conversely, in vehicle/motor-to-grid (V/M2G) mode, the system reverses power flow, allowing stored battery energy to be delivered back to the grid through the same DC link and BADC.

The battery is connected to the DC link via a bidirectional buck-boost converter, which manages voltage and current depending on the operating mode. During G2V (grid-to-vehicle) mode, the buck converter mode ensures controlled charging by stepping down the voltage as needed. During V2G (vehicle-to-grid), the boost converter raises the battery voltage to the appropriate level for grid injection. A coupling inductor, denoted as Lc, is used at the grid interface to reduce current ripple and support smooth energy exchange.

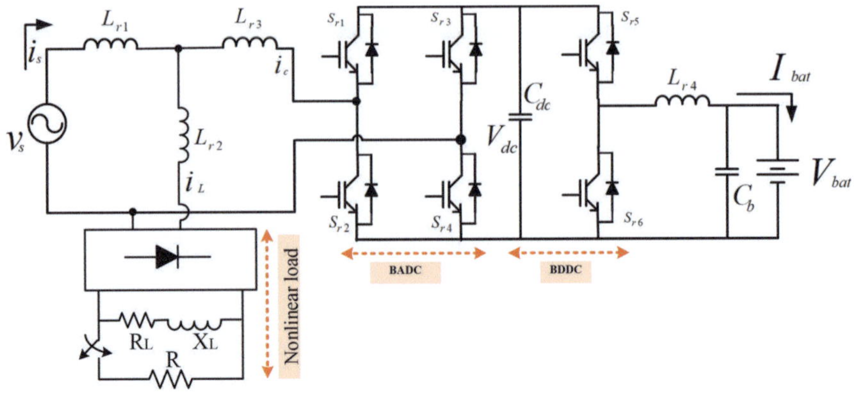

Fig. 1 Circuit diagram of the complete system

This architecture enables efficient bidirectional power flow while maintaining system stability and improving power quality, making it well suited for next-generation EV–grid integration.

3 Third-Order Sinusoidal Integrators Method [TOSSI]

Figure 2 illustrates the implementation of the Third-Order Sinusoidal Signal Integrator (TOSSI) algorithm used to extract the fundamental current component (i_f) from the load current (i_L). The algorithm separates i_L into two orthogonal components: the in-phase (i_α) and quadrature (i_β) signals. These components are essential for accurate current reference generation in G2M and M2G operations. TOSSI is tuned using g1 = 1.2, g2 = 2.5, and a damping factor ξ = 0.707 to ensure stability and effective filtering.

$$T_\alpha(s) = \frac{i_\alpha}{i_L} = \frac{g_1\omega^2 s}{s^3 + g_2\omega s^2 + (1 + g_1)\omega^2 s + g_2\omega^3} \tag{1}$$

$$T_\beta(s) = \frac{i_\beta}{i_L} = \frac{g_1\omega^3}{s^3 + g_2\omega s^2 + (1 + g_1)\omega^2 s + g_2\omega^3} \tag{2}$$

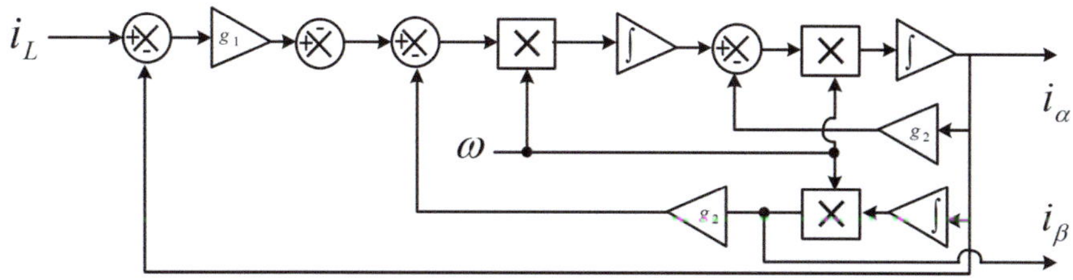

Fig. 2 TOSSI controller block diagram

4 Controller Design

Figure 3 depicts the control diagram for the proposed system. The EV charger is rated for 1.2 KW and has a battery current of ± 6A. The minimum AC voltage for power transmission is 110 V. As a result, a 200-V DC link capacitor is selected.

(A) BADC Control:

The fundamental component of the load current is extracted using the TOSSI filter. At the output of the TOSSI block, load current (i_L) is transformed into in-phase and quadrature-phase signals (I_α and I_β, respectively). Additionally, the amplitude of the fundamental current (i_f) is ascertained using Eq. 3:

$$i_f = \sqrt{i_\alpha^2 + i_\beta^2} \tag{3}$$

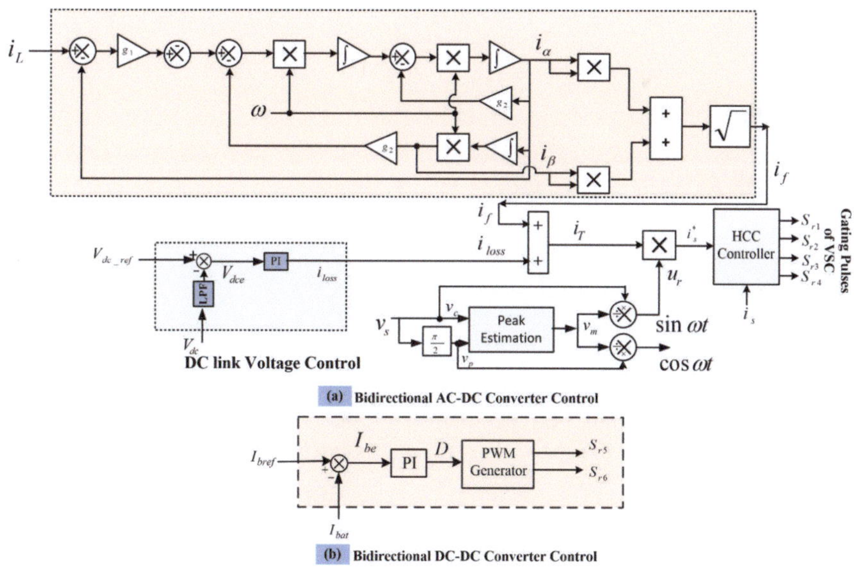

Fig. 3 Control structure of the proposed system

The DC link voltage (V_{dc}) is obtained depending on the reference voltage (V_{dc_ref}) and filtered using a low-pass filter to eliminate noise, thereby estimating the switching losses of the shunt compensator. A PI controller generates the switching loss component (I_{loss}) of the system by means of the difference between the actual and reference voltages, as stated in Equation (4). This dynamic reaction technique guarantees a correct loss estimate. The DC link voltage was maintained at 200V throughout the operation to guarantee consistent system performance and dependable switching loss measurement under different environments.

$$i_{loss} = i_l(n) + \left[K_p\{V_{dc}(n+1) - V_{dc}(n)\} + K_i\{V_{dc}(n+1)\}\right] \tag{4}$$

where the Kp, Ki is controller coefficient.

The magnitude of total reference current is obtained by adding the DC link loss component and fundamental load current component, as defined in Eq. 5:

$$i_T = i_f + i_{loss} \tag{5}$$

For the unit step generation, Eq. (6–8) has been used.

$$V_C = V_M \text{Sin}\omega t \tag{6}$$

$$V_P = V_M = \left(\text{Sin}\omega t + \frac{\pi}{2}\right) \tag{7}$$

Now, to calculate the peak voltage:

$$V_m = \sqrt{V_p^2 + V_c^2} = V_t \tag{8}$$

From Eq. (6–8) as

$$u(n) = \frac{V_c}{V_M} = \text{Sin}\omega t \tag{9}$$

Now, for the reference current generation, the unit step signal added by total reference current is given in Eq. (10):

$$i_s^* = i_T + u(n) \tag{10}$$

Ultimately, a hysteresis current controller (HCC) is employed for generating pulses for the regulation of BADC. Figure 3a illustrates the complete control scheme structure.

(B) BDDC Control:

To regulate the operation of the bidirectional DC–DC converter, an error signal is first generated by comparing the reference battery current (I_{bref}) with the actual battery current (I_{bat}). This error signal is then fed into a Proportional–Integral (PI) controller, as illustrated in Fig. 3b. The PI controller processes the error and produces a control signal, which is sent to a Pulse Width Modulation (PWM) generator. The PWM generator then generates the switching pulses required to control the converter. Specifically, the discharging mode is activated when switch S_{r6} is ON and S_{r5} is OFF, while the charging mode occurs when S_{r5} is ON and S_{r6} is OFF.

5 Result Analysis

Along with PQ enhancement, a MATLAB/Simulink environment is used to create and analyze the G2V/M and V/M2G modes. The section that follows provides a thorough explanation.

A. Operation in G2M/V mode:

Figure 4 presents the simulation results for the G2V (grid-to-vehicle) mode, which illustrates the voltage of grid (v_s), current of grid (i_s), load current (i_L), compensating current (i_c), DC link voltage (V_{dc}), battery voltage (V_{bat}), and battery current (I_{bat}). Since the v_s stays constant as the load increases at time t = 1.2 s, the i_s and i_L increase independently during the load change. Under both circumstances, there are slight fluctuations around V_{dc} = 200 V in the DC link. The battery rating chosen for the proposed charger is 96 V, 42Ah, and it is designed to run on a single phase, 110 V, 50 Hz supply. The grid supplies power to the battery and the battery charging current is negative, i.e., -6A. The signals v_s and i_s are in the same phase when in G2M mode (charging condition).Figure 5 shows the THD of the v_s, i_s, and i_L of THD 0.48%, 1.82%, and 17.69%, respectively, during G2M/V mode.

B. Operation in V/M2G Mode:

In Fig. 6, the performance under load change is shown during M2G mode. Grid voltage (v_s), grid current (i_s), load current (i_L), compensating current (i_c), DC link voltage (V_{dc}), battery voltage (V_{bat}), and battery current (I_{bat}) for M2G mode using MATLAB/Simulink are all shown in Fig. 6.

The current and grid voltage are seen to be out of phase with one another. During load change at t = 1.2 s, the v_s stays constant but the i_s and i_L increase independently. Under both circumstances, there are slight fluctuations in the DC link voltage around Vdc = 200 V, the reference value.

During both charging and discharging modes, the TOSSI control algorithm has been used to minimize the grid current's high ripple frequency. The THD of the v_s and i_s are 0.41% and 3.14%, respectively, as shown in Fig. 7. A nonlinear load in Fig. 7 has an i_L THD of 18.06%.

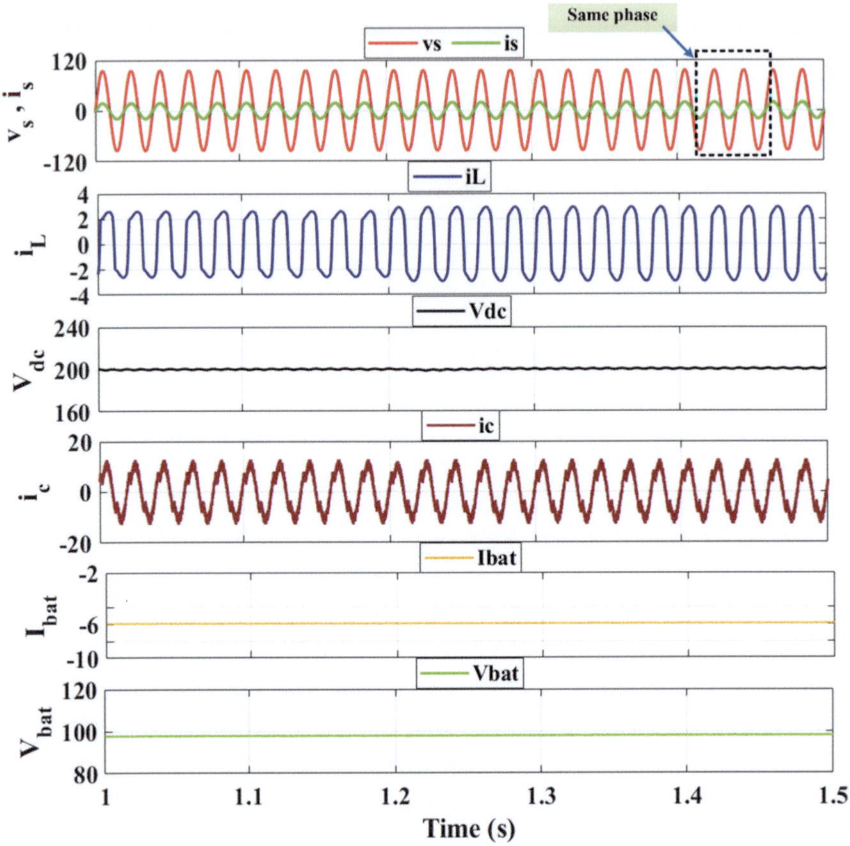

Fig. 4 Results during G2V mode, **a** Vs, is, **b** i_L, **c** V_{dc}, **d** i_c, **e** I_{bat}, **f** V_{bat}

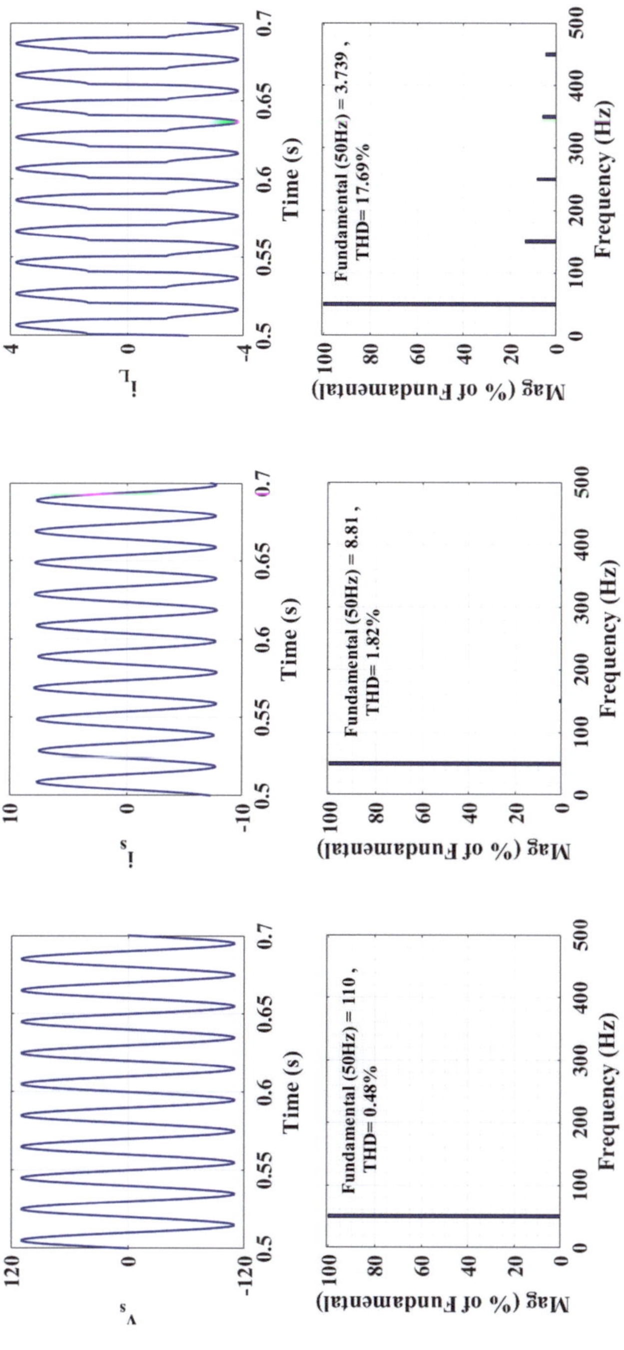

Fig. 5 THD analysis during G2V mode

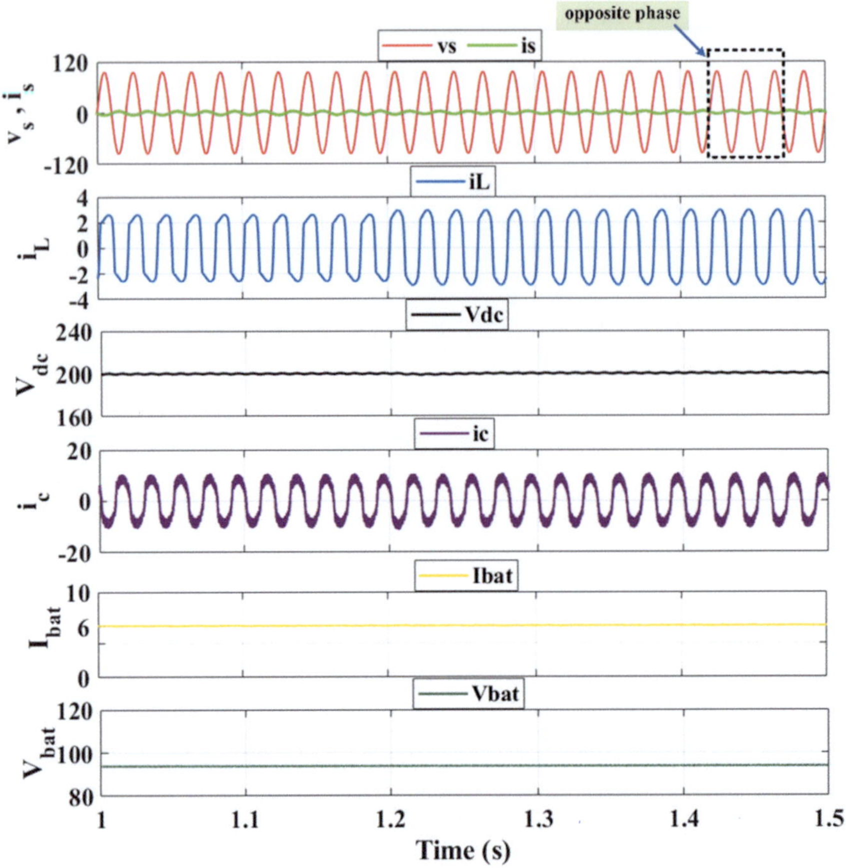

Fig. 6 Results during V2G mode **a** v_s, i_s, **b**i_L, **c** V_{dc}, **d** i_c, **e** I_{bat}, **f** V_{bat}

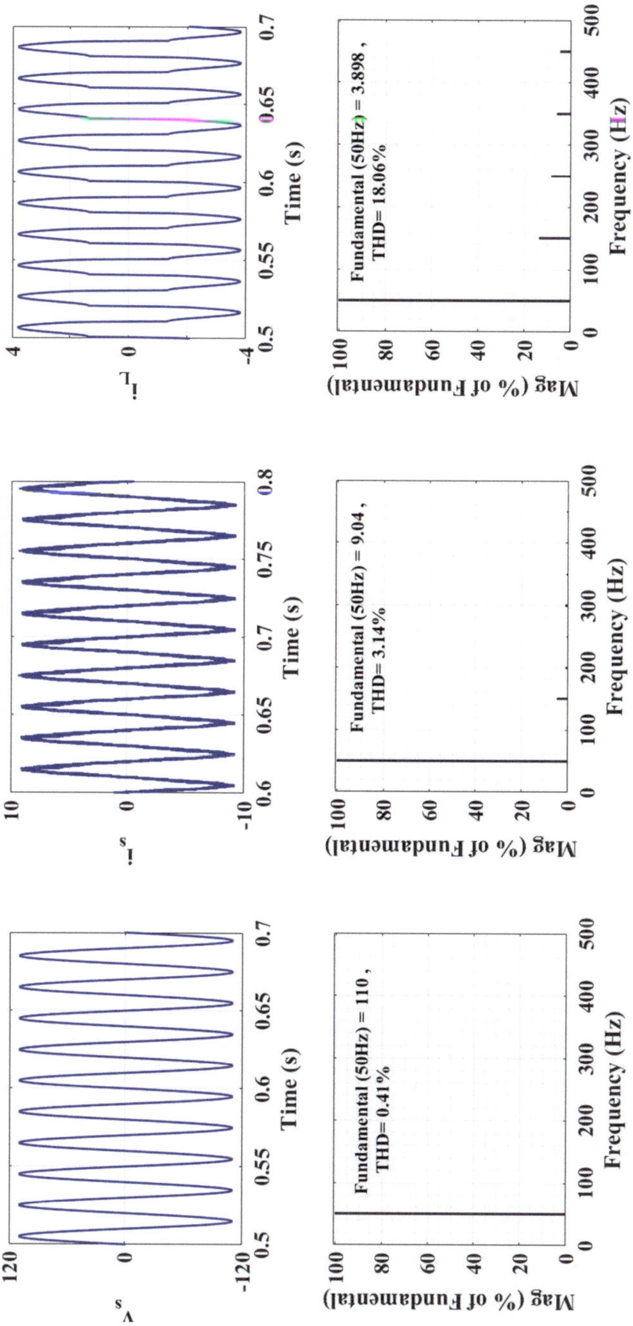

Fig. 7 THD analysis during V2G mode

6 Conclusion

In this research, the performance of two key components—a bidirectional AC–DC and DC–DC converter analyzed for application in electric vehicle (EV) charging systems supporting both G2M/V and V/M2G functionalities. The grid-side converter was effectively controlled using a Third-Order Sinusoidal Signal Integrator (TOSSI) controller. The obtained result confirms the feasibility as well as the effectiveness of the proposed control technique. Moreover, to develop and manage the power flow throughout the charging and discharging operation of the system, a bidirectional controller was successfully designed. Furthermore, the TOSSI controller demonstrated its ability to enhance power quality on the grid side. Its performance was validated through simulations under both steady-state and dynamic operating conditions, confirming its robustness and reliability for real-world EV–grid integration.

References

1. S Kundu M Singh AK Giri 2024 SPV-wind-BES-based islanded electrical supply system for remote applications with power quality enhancement Electr Eng 106 1 279 294
2. Nishant C et al (2023) Harmonic reduction strategies for EV charging in power systems using shunt active power filter. In: 2023 3rd International conference on emerging frontiers in electrical and electronic technologies (ICEFEET). IEEE
3. S Kundu M Singh AK Giri 2024 Synchronization and control of WECS-SPV-BSS-based distributed generation system using ICCF-PLL control approach Electr Power Syst Res 226 109919
4. Sudhanshu M, Singh A, Chittora P (2024) Design and development of reduced switch five level inverter for interfacing renewable energy sources and EV charger. Electr Eng 1–15
5. G Yadav M Singh 2024 Real-time investigation of grid-interactive EV charger with two-stage bidirectional converter under wide voltage range scenarios Electr Eng https://doi.org/10.1007/s00202-024-02609-z
6. Sombir K et al (2024) Robust and fast control approach for islanded microgrid system and EV charging station applications. Electr Eng 1–12
7. Sudhanshu M, Singh A, Chittora P (2023) Solar PV array based grid-connected bi-directional EV charger controlled using NARLMMN algorithm. In: 2023 9th IEEE India international conference on power electronics (IICPE). IEEE
8. S Kundu M Singh AK Giri 2024 Adaptive control approach-based isolated microgrid system with alleviating power quality problems Electr Power Compon Syst 52 7 1219 1234
9. Sombir K, Ashutosh K. Giri, Singh M, Kadiyan S (2024) Adaptive control strategy for isolated renewable energy-based generation system with intermittency. Electr Eng 1–13
10. Kundu S, Ashutosh K, Giri Sunil K (2024) An adaptive mixed-step size normalized least means fourth control approach for standalone power generation system considering dynamic conditions. IEEE J Emerg Sel Top Power Electron
11. S Mittal A Singh P Chittora 2024 Design and development of leaky least mean fourth control algorithm for single-phase grid-connected multilevel inverter Int J Circuit Theory Appl 52 1 328 345
12. Sudhanshu M, Singh A, Chittora P (2024) Power quality enhancement in single phase two level/five level converters using adaptive-RBFNN algorithm. Electr Eng 1–14
13. S Kundu M Singh AK Giri 2022 Control algorithm for coordinated operation of wind-solar microgrid standalone generation system Energy Sources, Part A: Recover, Util, Environ Eff 44 4 10024 10044

14. Amarendra P, Singh A, Mittal S (2025) Brain-inspired learning controller for grid-connected electric vehicle charging junction with improved power quality. Arab J Sci Eng 1–18
15. Sombir S, Singh M (2020) Voltage and frequency control of self excited induction generator integrated with PV system. IECON 2020. The 46th annual conference of the IEEE industrial electronics society. IEEE
16. Sudhanshu M, Singh A, Chittora P (2022) EV control in G2V and V2G modes using SOGI controller. In: 2022 IEEE 3rd global conference for advancement in technology (GCAT). IEEE
17. Ranjan A, Bhaskar DV, Kumar P, Muduli UR (2024) Comprehensive analysis of thermal effects on PMSM drive control in small commercial vehicles. In: 2024 IEEE international conference on power electronics, drives and energy systems (PEDES). IEEE, pp 1–6
18. Prakash A, Bhaskar DV, Ranjan A, Kumar P, Muduli UR (2024) Advanced control and modeling for enhanced BLDC motor efficiency and performance. In: 2024 IEEE international conference on power electronics, drives and energy systems (PEDES). IEEE, pp 1–6
19. Kumar P, Balanthi Beig AR, Khaled AJ, Devara VB, Behera RK, Muduli UR (2023) Continuous fast terminal sliding surface-based interrupt free operation of PMBLDCM Drive. In: IECON 2023–49th annual conference of the IEEE industrial electronics society. IEEE, pp 1–6
20. Agrawal A, Bhaskar DV, AlZaabi O, Kumar P, Al Hosani K, Panda M (2023) Hybrid dc boost converter for efficient voltage regulation and dimming control. In: 2023 IEEE IAS global conference on emerging technologies (GlobConET). IEEE, pp 1–6
21. Agrawal A, Bhaskar DV, AlZaabi O, Kumar P, Al Hosani K, Panda M (2023) Efficient hybrid boost converter: Enhancing voltage regulation for versatile lighting solutions. In: 2023 IEEE IAS global conference on emerging technologies (GlobConET). IEEE, pp 1–6
22. Siddiqui MK, Yadav G, Wasti AS, Ansari MA (2022) Performance analysis of power flow controller used for solar PV based grid integrated system. In: 2022 2nd International conference on emerging frontiers in electrical and electronic technologies (ICEFEET), Patna, India, pp 1–5. https://doi.org/10.1109/ICEFEET51821.2022.9847906
23. Mittal S, Alka S, Prakash C (2025) Rodrigues jacobi polynomial based control technique for on board EV charger in single phase power distribution system. Electr Power Syst Res 248: 111981
24. Kundu S, Giri AK, Kadiyan S, Singh S (2025) Power management and adaptive control method for a standalone distributed generation system under fault, DC-offset, and dynamic environments. Int J Ambient Energy 46(1). https://doi.org/10.1080/01430750.2025.2452245
25. Muzammal IM et al (2024) Improving reliability and stability of the power systems: A comprehensive review on the role of energy storage systems to enhance flexibility. IEEE Access
26. Singh S et al (2024) Design and control of electric vehicle charger using Bi-directional DC-DC CUK Converter. In: 2024 IEEE Third international conference on power electronics, intelligent control and energy systems (ICPEICES). IEEE
27. Singh N, Pratap B (2025) Robust observer design for variable speed wind turbine using a two-mass model. In: Proceedings of the international conference on systems, control and automation. ICSCA 2023. Lecture Notes in Electrical Engineering, vol 1260. Springer, Singapore. https://doi.org/10.1007/978-981-97-7384-8_15

Chapter 7
Electric and Hydrogen Vehicles: Systems, Infrastructure, and Future Trends

Kanchan Bala Rai

Abstract The global shift toward sustainable transportation is being propelled by the critical need to decrease carbon emissions, lessen dependence on fossil fuels, and tackle urban air pollution. As one of the primary sources of global greenhouse gas emissions, the transportation industry is experiencing a major transformation with the emergence of Electric Vehicles (EVs). This chapter offers a theoretical overview of the fundamental technologies facilitating this transition—from battery electric vehicles (BEVs) and fuel cell electric vehicles (FCEVs) to sophisticated charging systems, including wireless charging solutions. It examines essential technological elements such as power electronics, battery management systems (BMS), electric drive systems, vehicle control frameworks, and charging infrastructure. Particular focus is given to battery management systems for maintaining safety, extending longevity, and ensuring peak performance of EV batteries, alongside the importance of hydrogen fuel cells as a viable zero-emission energy alternative for extended-range and heavy-duty uses. The chapter also addresses vehicle-to-grid (V2G) technology, which allows for two-way energy exchange between EVs and the electricity grid, contributing to grid stability and energy adaptability. By presenting the theoretical principles and system-wide considerations of these enabling technologies, this chapter acts as a valuable reference for students, researchers, and professionals who seek to grasp and influence the advancement of electric mobility.

1 Introduction: Technologies for Electric Vehicle Applications

The world's shift toward sustainable mobility is being driven by the urgent need to reduce carbon emissions, reduce dependence on fossil fuels, and mitigate urban pollution. The transportation sector remains one of the most significant contributors to global carbon dioxide emissions, accounting for nearly 30% of total emissions

K. B. Rai (✉)
Ornate Solar, New Delhi, India
e-mail: kanchan@ornatesolar.in

S. Kundu et al. (eds.), *Hybrid Renewable Power Infrastructure for Sustainable Electric Vehicle Development*, Studies in Infrastructure and Control,
https://doi.org/10.1007/978-981-95-1323-9_7

in regions like Europe and the United States [1]. At the forefront of this shift is the rapid advancement of Electric Vehicle (EV) technology, which is revolutionizing the automotive sector. In contrast to conventional Internal Combustion Engine (ICE) cars, EVs rely on stored electrical energy in batteries to power high-efficiency electric motors, offering not only environmental benefits but also significant operating savings and a more pleasant driving experience [2].

India is especially well positioned at a strategic inflection point. With strong policy support through schemes like FAME-II, the PLI scheme, and state EV policies, the tide for EV adoption is gathering pace in the two-wheelers, three-wheelers, passenger cars, and commercial vehicles segments. It is underpinned by a thriving ecosystem of local innovation and production in key EV technologies.

This chapter addresses the enabling technologies at the core that enable electric mobility to be scalable and viable. They comprise power electronics, battery technology, electric motors, charging infrastructure, vehicle control systems, communication protocols, and BMS. All these technologies are essential to enable EVs to work, range, efficient, and safe.

Also, the coming together of Internet of Things (IoT), Artificial Intelligence (AI), and cloud-based telematics is converting EVs into smart, connected transportation systems. With India inching its way toward an environment-friendly car-centric future, understanding the tech fundamentals of EVs becomes vital for engineers, entrepreneurs, policymakers, and auto industry leaders alike.

As outlined in Table 1, battery technology—including lithium-ion and emerging solid-state batteries—sets the foundation for driving range, charging time, and overall efficiency. High-performance electric motors such as Brushless DC (BLDC), Permanent Magnet Synchronous Motors (PMSM), and induction motors convert electrical energy into mechanical power with minimal losses.

2 Battery Technology in EV Applications

Battery technology is at the core of electric vehicles (EVs), impacting virtually all performance attributes—range, speed, safety, cost, and efficiency. Batteries in EVs play the primary energy storage role of providing power to the electric motor and on-board electronics. Today, the most common battery type is the Lithium-Ion (Li-ion) battery, which is valued for its high energy-to-weight ratio, rapid charging, and long lifespan. Among Li-ion technologies, two major chemistries—LFP (lithium iron phosphate) and NMC (nickel manganese cobalt)—are widely used. LFP is known for its safety characteristics and higher cycle life, and hence it is suitable for Indian climatic conditions, whereas NMC is opted for because of its higher energy density, particularly in high-performance EVs.

An EV battery will consist of a cluster of cells arranged in modules, and the individual modules together form a battery pack with a Battery Management System (BMS) integrated into it. The BMS serves a vital purpose in safety through continuous

Table 1 Key technologies for electric vehicle applications

Technology area	Description	Importance
Battery technology	Lithium-ion batteries, solid-state technology, battery composition, and energy capacity	Establishes the range, charging duration, and overall performance of the vehicle
Electric motors	BLDC, PMSM, induction motors	Transforms electrical energy into mechanical power with high efficiency
Power electronics	Inverters, DC–DC converters, on-board chargers	Manages and transforms electrical energy effectively inside the vehicle
Battery management system (BMS)	Monitors cell voltage, temperature, SOC, and ensures safety	Enhances battery life, performance, and safety
Charging infrastructure	AC and DC chargers, fast chargers, swapping stations	Facilitates dependable and easy energy replenishment for electric vehicles
Vehicle control unit (VCU)	Embedded system for coordinating powertrain and auxiliary systems	Ensures smooth operation and driving behavior
Thermal management	Cooling/heating systems for battery and electronics	Maintains optimal temperature for performance and longevity
Connectivity and telematics	IoT, GPS, cloud-based diagnostics	Facilitates remote oversight, vehicle fleet administration, and anticipatory maintenance

cell temperature, voltage, and state of charge monitoring. Proper thermal management also assumes great importance in EVs to prevent overheating, provide the best performance, and extend battery life—especially across India's different climatic zones. Batteries will also have to comply with Indian safety standards like AIS-038 Rev.2 and AIS-156, with increasing usage of IP67-rated waterproof enclosures.

Vehicle range is directly impacted by energy density, a crucial component that establishes how much energy a battery can store per unit mass, as shown in Table 2. The number of charge/discharge cycles a battery can withstand before experiencing a noticeable drop in capacity is known as cycle life. The C-rate, which influences charging time and thermal load, shows how quickly a battery charges or discharges in relation to its capacity. Whether air-based or liquid-based systems, efficient thermal management guarantees that batteries run within safe temperature ranges to avoid deterioration or failure.

The cost, performance, safety, and use of electric vehicles (EVs) are all significantly influenced by the battery chemistry selection. Lithium Iron Phosphate (LFP) batteries are used in two-wheelers, three-wheelers, and low-cost city cars like those made by Tata, Ola, and Hero because of their high thermal stability, long cycle life,

Table 2 Key features of EV batteries

Parameter	Explanation
Energy density	Determines how much energy can be stored per kg—higher = more range
Cycle life	Number of charge/discharge cycles before capacity drops
C-rate	Charging/discharging speed relative to battery capacity
Thermal management	Maintains safe temperature; air- or liquid-based systems are used
Battery Management System (BMS)	Ensures battery health, controls charging, prevents overheat/explode
State of Charge (SoC)	Real-time indication of battery percentage

Table 3 Popular EV battery chemistries

Type	Advantages	Disadvantages	Applications
LFP	High thermal safety, long life, low cost	Lower energy density	2W, 3W, city cars (Tata, Ola, Hero)
NMC	High energy density, compact	Expensive, thermally less stable	Premium EVs (MG ZS, Hyundai Kona)
Solid-State	Ultra safe, fast charge (future tech)	High cost, under development	R&D stage

and affordability, as shown in Table 3. However, their use in long-range EVs is limited by their lower energy density.

Key Challenges in India

1. High Cost of Imported Cells: India imports ~ 90% of battery cells.
2. Temperature Extremes: Affect battery life and performance.
3. Battery Recycling Ecosystem: Still in nascent stage.
4. Charging Time versus Infrastructure: Lack of fast-charging network impacts adoption.

3 Power Electronics in Electric Vehicle Applications

Power electronics is the key control system for electric vehicles (EVs), controlling and converting electrical energy among the battery, electric motor, and other auxiliary systems. Without power electronics, the operation of the vehicle would be impaired, as it plays a critical role in controlling the energy flow and guaranteeing smooth operation. It is crucial for efficiency, safety, and driving performance.

The three primary components of power electronics in an EV include the following:

1. Inverter.
2. DC–DC converter.
3. On-Board Charger (OBC).

1. Inverter:

The inverter transforms the DC (direct current) electricity from the battery into the AC (alternating current) needed to operate the electric motor. It also manages regenerative braking, where the AC power generated by the motor during braking is converted back into DC and stored in the battery. Modern inverters utilize IGBTs or SiC-based MOSFETs for effective high-frequency switching. The control algorithms of the inverter (such as field-oriented control—FOC) ensure smooth control of torque and speed. Efficiency of EV battery charging primarily depends on the power electronic converter topologies used in the chargers. Converter topologies presented in [3] use single-stage AC–DC power conversion for EV battery charging. Two-stage conversion systems use an AC–DC converter followed by an active power factor correction (PFC) and a DC–DC converter for tight regulation of output voltage against line and load fluctuations. Unidirectional power converters are mostly used for charging the batteries of small EVs and interfacing renewable energy sources to the charging systems [4, 5].

Bidirectional power converters can provide both the features of EV charging from the grid, i.e., grid-to-vehicle (G2V) mode operation and feeding the battery-stored electrical energy back to the grid, i.e., vehicle-to-grid (V2G) mode operation [6]. However, EV battery charging systems, being high power load, may sometimes have the adverse effects of protection relay tripping and overloading of distribution transformers, fuses, cables, etc., if the charging is done at peak hours. Charging at off-peak hours and proper coordination among different charging stations are therefore recommended to be adopted to minimize the adverse effects of load imbalance, energy shortage, voltage instability, power quality degradation, etc. [7].

2. DC–DC Converter:

DC–DC converters are used to provide a constant output voltage from a variable or an oscillating input DC voltage. DC–DC converters come in two categories, linear regulators and switch converters. The output current in a linear regulator is provided directly from the power supply, and therefore performance is given as ratio of the output to input voltages. It goes without saying that, lower efficiency is obtained, if the output voltage is much lower than the supply voltage. As a result of intercepted energy transmission, switching converters are more powerful than linear regulators. This is done by repeatedly exchanging components from energy storage to transfer part of power from the input source to the load end. DC–DC converters can be categorized into two wide classes (except resonant converters): capacitive and inductive. Two serious demerits in inductor-based DC–DC converters are problematized as follows: one is a result of high voltage spikes occurring in such converters, which must be either damped or repaired, while the rest of the circuit will be damaged, but power electronic switches unable for these constraints will explode. The other problem of inductive converters is that the input current is pulsating in the natural world in these DC–DC converters, which can generate EMI from other lines of equipment and conductors. DC–DC converters with capacitors are referred to as capacitive converters, which are very much used in low power applications and do not involve any isolation. These are relatively minimum noise, almost radiated EMI-free, and very easily fabricated into ICs, making these kinds of DC–DC converters are very much utilized in mobile device power source management and Electric Vehicles (EVs). In EVs, switched capacitor converter [8] is crucial for managing the speed of the vehicle and powering the electronics loads like LEDs, head lamps, and audio systems.

The battery pack delivers high-voltage DC ranging from 200 to 800 V; however, electronic components in electric vehicles, such as lights, infotainment, wipers, and battery management systems, operate at either 12 V or 48 V. A DC–DC converter reduces the high voltage to the necessary lower voltage, which allows for the safe and efficient functioning of auxiliary systems. In hybrid systems or hybrid EV, bi-directional converters are utilized for both increasing and decreasing voltage control [9].

3. On-Board Charger (OBC):

The On-Board Charger (OBC) transforms alternating current (AC) from the charging station into direct current (DC) to recharge the battery. It is engineered to accommodate different input voltages and various charging standards such as Bharat EV, CCS2, and CHAdeMO. In India, the AC type-2 and DC CCS2 protocols are followed under the FAME-II initiative. High-efficiency OBCs minimize charging duration while ensuring the safety of the battery. Various techniques are employed for battery controlling which enhance the efficiency and lifetime span. But conventional battery controller employs proportional integral controller which leads to stability issues [10]. For the charging of the batteries, we need a charger which is also referred to as the on-board charger. This charger's purpose is to harness the AC supply of power from the home or any other power source which typically is 240 V. Then this AC power supply is then transformed into a DC supply so that it charges the battery. It also monitors the voltage, current, temperature, and all other parameters of the battery. There is yet another component named as the power electronics controller which assists in the transmission of the power supply from the battery and speed and torque control of the vehicle. The system that is responsible for managing or controlling the temperature of some components of the vehicle is also fitted along with the rest of the components, and such a system is referred to as the thermal system.

Figure 1 illustrates a power conversion system with two stages. The totem pole converter (left) is a Power Factor Correction (PFC) circuit, which converts 230 V, 50 Hz AC to regulated high-voltage DC with high efficiency and near-unity power factor. It employs fast-switching transistors (Q1–Q4) and an input inductor (L_1) to condition the input current. The second phase is an isolated full-bridge converter, employing transistors (Q5–Q8), a high-frequency transformer, and synchronous

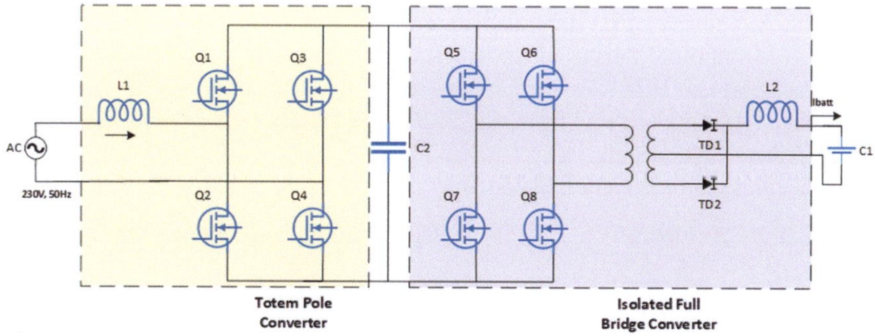

Fig. 1 Totem pole PFC with isolated full-bridge DC–DC converter for high-efficiency battery charging

rectification (TD_1 and TD_2) to achieve galvanic isolation and step down the DC voltage. The output inductor (L_2) and capacitor (C_1) filter the current to supply a smooth charging current (I_{batt}) to the battery.

4 Battery Management System (BMS) for EV Applications

A Battery Management System (BMS) is the key electronic device that controls electric vehicle (EV) batteries, maintaining safety, longevity, and peak performance. In India's growing EV ecosystem, especially under the FAME-II and National Electric Mobility Mission (NEMMP) programs, a well-designed BMS has emerged as a key technical specification for EV certification.

A BMS mainly carries out four primary functions:

1. **Monitoring:** It continuously monitors cell voltages, currents, and temperatures to assess the State of Charge (SOC) and State of Health (SOH) of every single cell and the whole battery pack.
2. **Protection:** It safeguards the battery against overvoltage, undervoltage, overcurrent, short circuit, and thermal runaway conditions through triggering shutdowns or alarm.
3. **Balancing:** Because of inevitable manufacturing differences, cell voltages can drift apart over time. The BMS ensures cell balancing—either actively (moving energy) or passively (draining excess energy)—to maximize usable battery capacity.
4. **Communication:** The BMS communicates with the vehicle's Energy Management System (EMS) and charger through CAN or UART protocols, following ISO 15118 or OCPP standards in more sophisticated systems.

In the context of India, wherein temperature variation is high and road conditions are problematic, it becomes necessary to have thermal management built into the BMS. A majority of EV fires that were reported in India (2021–22) have been attributed to improper design of the BMS. Because of this, MoRTH (Ministry of Road Transport and Highways) regulations today mandate compliance to functional safety (like AIS-156, adopted from ISO 26262) for battery packs, with the inclusion of an effective BMS design.

A typical BMS architecture includes

1. **Sensing Circuitry:** For accurate voltage, current, and temperature measurements.
2. **Microcontroller Unit (MCU):** For running algorithms such as Kalman filters or Coulomb counting for SOC/SOH estimation.
3. **Balancing Circuit:** Using MOSFETs or active balancing ICs.
4. **Communication Interface:** For diagnostics and real-time monitoring.

Figure 2 depicts the monitoring and management system of a lithium-ion battery, which is vital for safe and efficient operation, particularly in electric vehicle and energy storage applications. The battery is constantly monitored for essential parameters like bus voltage, current, temperature, and cell voltage that flow into a Battery Management System (BMS). The BMS carries out a number of key functions such as monitoring battery cells, SOC estimation, SOH estimation, and RUL estimation. It also computes the SOE and thermal conditions to avoid overheating. The system also offers fault diagnosis to identify and react to abnormalities, and all the data related to it is processed through data acquisition and storage systems for analysis and system optimization. This overall structure guarantees the consistency, safety, and durability of lithium-ion battery systems [11].

Recent advancements include the adoption of AI-powered predictive algorithms in battery management systems (BMS) to predict failures and cloud-enabled BMS platforms (IoT-based) that allow fleet managers to track health remotely, a function that is growing particularly critical for India's electric vehicle fleets as well as ride-sharing

Fig. 2 Battery management system (BMS)

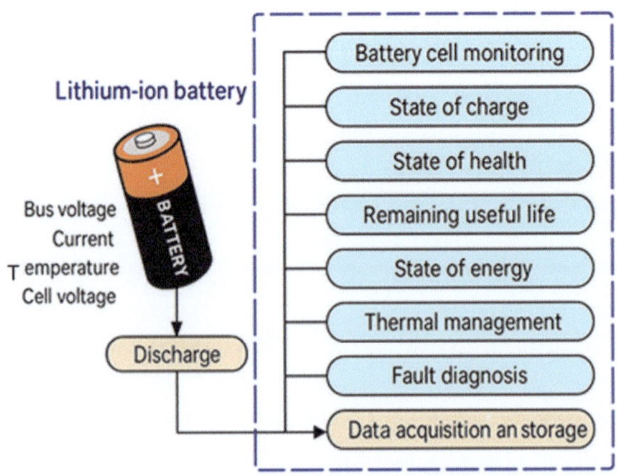

operations. In short, a well-designed BMS not only protects the battery but also optimizes performance, enhances user confidence, and supports regulatory compliance. For Indian e-vehicle manufacturers, the investment in strong BMS capabilities is not only an option—it is the key to commercial success and vehicle safety.

4.1 Mathematical Modeling in Battery Management Systems (BMS) for EV Applications

Effective functioning of a Battery Management System (BMS) demands proper estimation of vital battery parameters like State of Charge (SOC) and State of Health (SOH). Mathematical models are essential to predict, control, and optimize battery behavior in real time.

4.1.1 State of Charge (SOC) Estimation

State of Charge (SOC) is the remaining charge in a battery with respect to its capacity. This is one of the key parameters used for the estimation of driving range and for the safe operation of electric vehicles. Coulomb counting is the simplest and most widely used method for SOC estimation and can be described mathematically as follows:

$$SOC(t) = SOC(t_0) - \frac{1}{C_{nominal}} \int_{t_0}^{t} I(\tau)d\tau \tag{7.1}$$

where

- $SOC(t)$ is the state of charge at time t,
- $SOC(t_0)$ is the initial state of charge at time t_0,
- $C_{nominal}$ is the nominal capacity of the battery (Ampere-hours, Ah),
- $I(\tau)$ is the instantaneous battery current at time τ, positive during discharge, negative during charging.

Explanation

The Battery Management System (BMS) tracks the current entering and leaving the battery over a period. By accumulating the current, it determines the overall change in charge. The State of Charge (SOC) decreases when the vehicle is in use and rises during charging phases. Although this approach is straightforward, it is vulnerable to inaccuracies from sensors and noise, leading more sophisticated BMS to include correction techniques like voltage-based adjustments and Kalman filtering methods. If an EV battery with a nominal capacity of 50 Ah starts at 100% SOC, and a constant discharge current of 10 A is drawn for 1 h, the new SOC can be calculated as follows:

$$SOC = 100\% - \left(\frac{10 \times 1}{50}\right) \times 100 = 80\%$$

Thus, after 1 hour of discharge at 10 A, the SOC drops to 80%.

4.2 Regulatory Relevance

In India, according to AIS-156 and Automotive Industry Standards, the Battery Management System (BMS) must accurately estimate the State of Charge (SOC) within specified error limits to comply with safety, warranty, and eligibility criteria for subsidies under FAME-II regulations. Inaccuracies in SOC can result in over-discharge, thermal incidents, or concerns about vehicle range, which are all vital challenges for the acceptance of EVs in India.

Due to internal resistance and efficiency losses, the SOC exhibits a characteristic profile during charge and discharge cycles, showing a nonlinear increase during charging and a nearly linear decrease during discharge, as shown in Fig. 3a. Moreover, Fig. 3b indicates the battery's thermal response, demonstrating a visible temperature increase with extended use that could affect lifespan and performance if improperly controlled.

5 Battery Electric Vehicles (BEVs) and Lithium-Ion Batteries

Battery Electric Vehicles (BEVs) are transforming the landscape of transportation by providing a clean and efficient option compared to internal combustion engine (ICE) vehicles. Unlike hybrids, BEVs function solely on electric power stored in large,

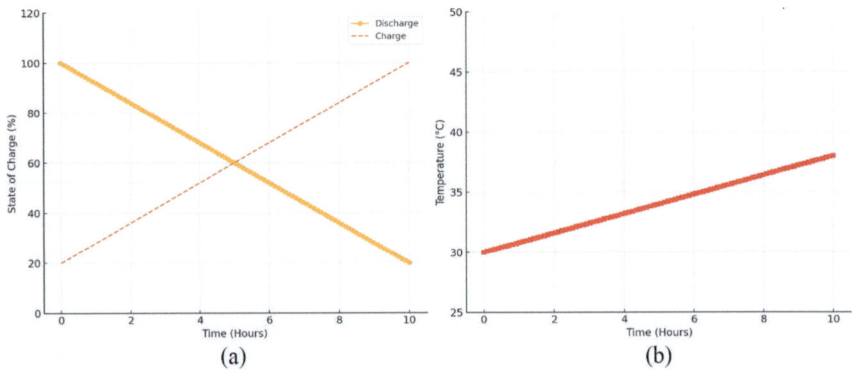

Fig. 3 **a** SOC variation during charge and discharge, **b** Battery temperature rise during operation

rechargeable battery units. At the center of this transition is the lithium-ion battery, which has become popular through its superior features compared to conventional forms of storage such as lead-acid or nickel-metal hydride batteries.

Lithium-ion batteries offer a number of significant advantages to BEVs. They provide great energy density, so that cars are able to store large quantities of energy in relatively lightweight and compact battery packs [12]. This translates directly into longer driving range—a critical factor in consumer acceptance. Contemporary BEVs will typically range from 250 to more than 500 km per charge, depending on battery size and car design. Another important advantage is the quick charging capability of lithium-ion batteries. Improvements in battery chemistry and advances in fast-charging infrastructure have enabled electric vehicles to be charged 80% of their battery capacity in as little as 30 to 60 min, significantly reducing the time-related inconveniences normally associated with EVs.

In addition to the advantages of range and charging time, lithium-ion batteries also provide a longer life of operation. New battery chemistries like Nickel Manganese Cobalt (NMC) and Lithium Iron Phosphate (LFP) have improved the lifecycle performance of batteries, which are capable of lasting 8 to 10 years or more under proper usage and environmental conditions. Their light weight also helps improve vehicle performance, benefiting acceleration, handling, and overall efficiency. Moreover, BEVs powered by lithium-ion batteries have lower maintenance than conventional vehicles because they have fewer moving mechanical parts, thus incurring lower maintenance costs over the vehicle's lifetime.

However, lithium-ion batteries require attentive monitoring to ensure safety, longevity, and maximum performance. Battery Management Systems (BMS) are embedded in BEVs to constantly monitor key parameters including cell voltages, temperatures, and currents. The difficulties for LEV batteries are that the chargers must be adapted to accommodate the battery voltage. There is no common value of battery pack voltage, and it may be between 48 and 72 V. The charging standards adopted for EVs are made for battery packs with high voltages. Therefore, those standards can't be used for LEV batteries. Since LEV batteries lack active cooling systems, charging poses both shock and fire risk. In China, there have been some instances of fire due to charging of LEV batteries within flats. This prompted a policy response to dissuade home charging and encourage public charging with safety controls.

Safety standards EN 50,604 and ISO 18243 have been attempted to be developed. The following sub-section discusses battery parameters in detail [13]. The BMS protects the battery by averting dangerous conditions such as overcharging, deep discharging, and excessive heat, in addition to balancing voltage levels between cells to achieve even usage and health across the battery pack. This is especially important in Indian climates, where temperatures can rise up to over 45 degrees Celsius in summer. Thermal management systems, typically including sophisticated liquid cooling or active air-cooling technology, are hence integrated with battery packs to ensure safe and efficient temperatures of usually between 20 °C and 40 °C.

The importance of proper battery management has led to the introduction of stringent regulatory standards in India. Specifications like AIS-156, as per international safety standards, govern the design, testing, and certification of EV battery packs and management systems. Complying with these standards is crucial for manufacturers who wish to offer safe and reliable BEVs in the Indian market.

In the future, India is also actively promoting electric mobility by government policies such as the Faster Adoption and Manufacturing of Hybrid and Electric Vehicles in India (FAME-II) scheme and the Production Linked Incentive (PLI) scheme for Advanced Chemistry Cell (ACC) battery production. These efforts would promote a strong indigenous battery manufacturing ecosystem, reduce imports, and bring EVs within the reach of Indian consumers. LFP batteries are increasingly popular in the Indian market for electric two-wheelers and three-wheelers based on their thermal stability and economic advantage, while NMC batteries are widely used in premium electric four-wheelers where energy density and extended range are the most important factors. The ongoing expansion of battery-swapping infrastructure, coupled with the swift rollout of fast-charging points, is mitigating range and charging convenience woes, especially critical for high-density and expansive geographies such as India. Major players in the industry like Tata Motors, Mahindra Electric, Ather Energy, and Ola Electric are focusing big bucks on R&D efforts to create battery packs specific for Indian road conditions, climate, and driving behaviors.

Over the coming years, advancements in battery technology, including solid-state and lithium-sulfur batteries, are predicted to make battery electric vehicles (BEVs) more efficient, cost-effective, and eco-friendly. Solid-state batteries particularly hold the promise of higher energy densities, faster charging times, and significantly improved safety in comparison to traditional lithium-ion designs. India has set ambitious goals for the coming years, with a goal of 30% adoption of electric vehicles by the year 2030. Achieving this vision will require continued improvements in battery technology, local manufacturing, efficient battery management systems, and robustly supported charging infrastructure. Lithium-ion technology will remain at the core of this transformation, driving India's mobility revolution toward cleaner, smarter, and greener road transportation.

6 Fuel Cell Electric Vehicles (FCEVs) and Hydrogen Technology

Fuel Cell Electric Vehicles (FCEVs) offer a revolutionary means of attaining zero-emission transportation, operating on a different mechanism than Battery Electric Vehicles (BEVs). Instead of storing electrical energy, FCEVs generate electricity internally by combining hydrogen gas with atmospheric oxygen in a fuel cell stack. The electrochemical process produces only water vapor as a byproduct, which makes FCEVs an extremely environmentally friendly option for transportation. With the insistence of tough rules pertaining to global warming, fuel economy, and traditional energy constraints, electric, hybrid, and FC cars have been attending to notice from car-makers and authorities [14]. In particular, proton exchange membrane (PEM)

[15]-based FCHEV is being attracted in this decade because of growing concern about pollution and associated problems toward environment [16]. Although powertrain development is challenging in FC vehicle and on-board hydrogen tank to be kept is challenging. At the end of nineteenth century, greenhouse gas effects prevail, which are prime for research community to decide on the emission of such gases with governments' concern. The governments have made decision to reduce the thermal plants, application of ICEs and other such polluting factors in the Kyoto policy. The FC vehicles emit no emission and are longer lasting.

Figure 4 illustrates the powertrain architecture of a fuel cell electric vehicle (FCEV), highlighting the integration of key subsystems for efficient propulsion. The fuel tank stores hydrogen, which is supplied to the fuel processor and then to the fuel cell (FC) subsystem, where electrochemical reactions generate electrical power. This power is regulated and conditioned through a bidirectional DC–DC converter [17], which facilitates power exchange between the fuel cell and other energy storage components, such as a battery (not shown explicitly). The conditioned DC power is converted to AC by the motor inverter and used to drive the traction motor, which transmits mechanical power to the wheels through the transmission. This configuration enables dynamic load balancing, regenerative braking, and stable voltage regulation for improved efficiency, range, and reliability of the FCEV system.

A bidirectional DC–DC converter in a fuel cell electric vehicle enables power flow between the fuel cell and battery, supporting regenerative braking and efficient energy use. It stabilizes voltage levels and manages load sharing to enhance fuel cell lifespan. This improves overall system efficiency, drivability, and extends vehicle range.

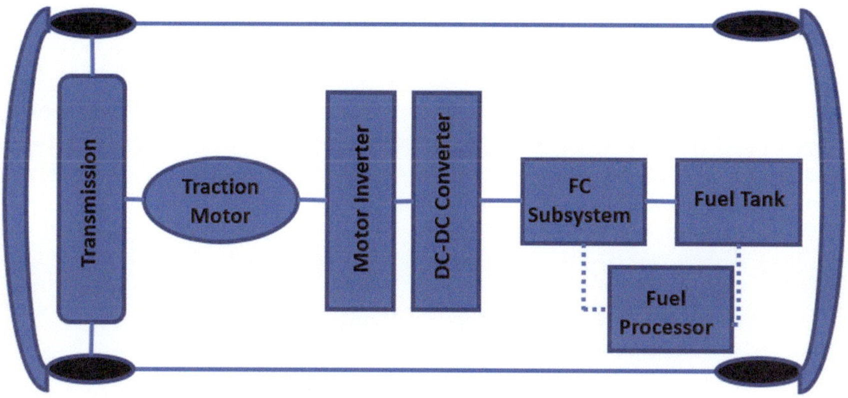

Fig. 4 Block diagram of fuel cell electric vehicle (FCEV) powertrain system

6.1 Advantages of FCEVs

One of the strongest points of FCEVs is that they have very long driving distances. Existing FCEVs tend to cover from 500 to 700 km on one tank of hydrogen and, therefore, are specially suited for long-distance driving and business applications. Moreover, refueling an FCEV is extremely rapid and usually is not more than five minutes, similar to standard gasoline or diesel cars. This quick refueling capability bypasses one major weakness that has come with BEVs—slow charging times—hence being more convenient for customers who are accustomed to conventional refueling procedures. At the heart of an FCEV is the hydrogen fuel cell stack, which splits hydrogen molecules into protons and electrons. The electrons travel through an external circuit, generating electricity to drive the vehicle's engine, while the protons combine with oxygen ions to form water. This process is very efficient, quiet, and clean, with no tailpipe emissions except for pure water vapor.

6.2 Challenges in Hydrogen Storage and Infrastructure

Despite having outstanding advantages, fuel cell electric vehicles (FCEVs) face insurmountable barriers that have checked their wide implementation, particularly in developing nations like India. The storage of hydrogen is specially intricate, with hydrogen gas being highly light in nature and needs to be stored under very high pressure (commonly around 700 bar) or at very low temperatures in order to effectively employ them in cars. Designing storage tanks that are safe, lightweight, and robust under such severe conditions greatly adds to the cost and engineering complexities of FCEVs [18]. Even more serious is the lack of hydrogen refueling infrastructure. Creating a hydrogen refueling station is much more expensive and complicated than building an electric vehicle charging station. Every hydrogen station requires special high-pressure equipment, rigorous safety protocols, and significant investment. In India, hardly any hydrogen refueling stations exist for personal vehicles, and the hydrogen infrastructure is in its infancy stage.

6.3 Future Outlook in India

The success of Fuel Cell Electric Vehicles (FCEVs) depends to a large extent on the source of hydrogen. Currently, most of the hydrogen produced worldwide is derived from fossil fuels, or gray hydrogen, which reduces the environmental benefits of FCEVs. To realize the complete potential of hydrogen fuel cars, India must make green hydrogen, produced by electrolysis powered by renewable sources like solar power and wind energy, its priority. Realizing this, the Government of India has launched the National Hydrogen Mission with the goal to make India a hub for the manufacturing and consumption of green hydrogen. Early projects with hydrogen-powered buses and heavy trucks are already underway in different parts of the country. These early efforts demonstrate that, while challenges exist, well-targeted funding and strong policy support can make the successful integration of FCEVs into India's overall transportation system possible.

6.4 Conclusion

Fuel Cell Electric Vehicles (FCEVs) offer the advantages of extended driving ranges and quick refueling times, making them suitable for long-distance travel, commercial transport, and heavy-duty vehicles. However, for universal adoption, technological challenges related to hydrogen storage need to be overcome, infrastructure needs to be enlarged, and green hydrogen production ensured. With combined industry–government efforts, FCEVs can play a big role toward establishing a sustainable and diverse electric transportation system for India.

6.5 Mathematical Model for Hydrogen Fuel Cell Operation

The basic electrochemical reaction in a hydrogen fuel cell [19] is

$$H_2 + \frac{1}{2}O_2 \rightarrow H_2O + ElectricalEnergy \tag{7.2}$$

6.5.1 Fuel Cell Voltage (Nernst Equation)

The ideal **open-circuit voltage** (no current drawn) of a hydrogen fuel cell can be calculated using the **Nernst equation**:

$$E = E^0 + \frac{RT}{2F} ln\left(\frac{P_{H_2} \times P_{O_2}^{0.5}}{P_{H_2O}}\right) \tag{7.3}$$

where

- E = Actual reversible voltage (V),
- E^0 = Standard reversible voltage at 25 °C (typically ≈ 1.23 V),
- R = Universal gas constant (8.314 J/mol),
- T = Absolute temperature (K),
- F = Faraday constant (96,485 C/mol),

- P_{H_2}, P_{O_2},P_{H_2O} = Partial pressures of hydrogen, oxygen, and water vapor, respectively.

6.6 Stack Output Voltage

Since practical fuel cells suffer from losses (activation, ohmic, concentration), the **actual cell voltage** is

$$V_{cell} = E - V_{activation} - V_{ohmic} - V_{concentration} \tag{7.4}$$

where

- $V_{activation}$ = Loss due to electrode kinetics,
- V_{ohmic} = Loss due to internal resistance,
- $V_{concentration}$ = Loss due to mass transport limitations.

Power output of fuel cell stack:

If V_{cell} is the cell voltage and I is the current, the fuel cell output power is

$$P_{stack} = N_{cells} \times V_{cell} \times I \tag{7.5}$$

where

P_{stack}= Total electrical power output (W).
N_{cells} = Number of individual cells in the fuel cell stack.

6.7 Emerging Technologies Enhancing Electric Vehicle (EV) Usability

6.7.1 Introduction

The global shift toward electric vehicles (EVs) marks a significant transformation in both the automotive and energy industries. Although the initial adoption of EVs encountered obstacles such as range anxiety, lengthy charging times, and insufficient infrastructure, recent advancements in technology are significantly enhancing the usability and convenience of electric vehicles. Innovations in fast-charging networks, wireless charging technologies, and vehicle-to-grid (V2G) capabilities are changing the way users engage with electric vehicles, offering the potential to make EVs a viable and appealing option for widespread use. This chapter examines these emerging trends, their current development stage, and the anticipated influence on the future of mobility, particularly within the context of India.

6.8 *Fast-Charging Networks: Revolutionizing Range Anxiety*

6.8.1 Understanding Fast Charging

Rapid charging, often known as DC fast charging or Level 3 charging, greatly shortens the duration needed to charge an electric vehicle (EV) when compared to conventional AC charging (Level 1 and Level 2). By utilizing direct current (DC) at elevated voltage and current levels, fast chargers can restore an EV's battery from 20 to 80% in just around 30 min.

AC charging system utilizes AC–DC converter to charge EV batteries at various power levels. Level 1 (110 V, 12–16 A, 1.5 kW) chargers are commonly adopted to recharge the batteries of small EVs in a time span between 0.5 and 12.5 h. These are largely preferred for overnight charging [20, 21].

Level 2 (240 V, 16–80A, 14.4 kW) charging systems are employed for EVs with on-board charger and the batteries are charged by direct connection to the utility grid [22]. Level 3 (240/480/600 V, 150–400A, > 14.4 kW) charging systems employ permanently wired chargers, solely for EV battery charging. Level 3 chargers are typically fast chargers and do not exceed 30 min to charge EV batteries. DC chargers are installed with dedicated wiring and require mounting, typically in charging stations, and tend to charge faster than AC chargers. Only DC supply voltage must be regulated for such systems to meet the battery requirement. DC chargers can also be categorized under the levels of Level 1 (200–450 V, 80A, 36 kW), Level 2 (200–450 V, 200A, 90 kW), and Level 3 (200–600 V, 400A, 240 kW).

6.8.2 Global Development of Fast-Charging Infrastructure

Countries like the United States, China, and several European nations have invested heavily in building large fast-charging networks. Some of these examples of public–private sector collaboration include Tesla's Supercharger network, Electrify America, and China's State Grid fast-charging stations, which are spearheading infrastructure development. Technological advancements have also increased charger output, such that ultra-rapid chargers can deliver over 350 kW, enabling a car like the Porsche Taycan to achieve 80% charge in about 20 min.

6.8.3 Fast-Charging Initiatives in India

India has acknowledged the significance of fast-charging infrastructure and initiated several programs:

The government is backing the deployment of thousands of fast chargers across the country under the FAME-II (Faster Adoption and Manufacturing of Hybrid and Electric Vehicles) initiative. Energy Efficiency Services Limited (EESL) along with firms such as Tata Power, Fortum, and Ather Energy is committing resources to develop fast-charging stations, especially along highways and in urban areas.

Standards such as Bharat DC-001 and Bharat AC-001 have been tailored specifically to Indian conditions, taking into account temperature variations and grid reliability. By the year 2030, India aims to create a comprehensive network of fast chargers to facilitate intercity travel and encourage the adoption of electric vehicles in various segments.

7 Wireless Charging: Future of Effortless Refueling

7.1 Basics of Wireless Charging

Wireless charging, referred to as inductive charging, transmits energy from one coil to another (the transmitter and the receiver) through electromagnetic fields, eliminating the need for physical connections. This system generally comprises a charging pad on the ground and a receiving pad on the vehicle [22, 23]. Nikola Tesla invented power transfer by wireless mode. It doesn't have physical electrical connection between power supply and EVs. Power was transferred from primary to secondary side is very cumbersome job and that will be addressed by researchers. In year of 1999, distance between primary and secondary winding was only 3 mm and power was transferred up to 8.3 kW with efficiency of 97%. WCS classification is explained above in various manners like static, dynamic, and quasi-dynamic wireless charging scheme, and near- and far-field charging infrastructure. Near field is classified as inductive contactless charging, resonant inductive contactless charging (RICC), capacitive contactless charging (CCC), magnetic resonance contactless power transfer (MRCPT), permanent magnetic gear coupling contactless power transfer, and for far-field laser and microwave wireless charging. Inductive contactless charging scheme is one of the most ancient schemes of battery charging. In the year of 1914, Nikola Tesla developed how can power transferred from one side to another side without connection. According to survey inductive contactless charging scheme, it has tested vast range of power transfer in milliwatts to kilowatts in EVs application. In 1996, General Motors launched Chevrolet S10 eV which was equipped with AC EVSE Level II and AC EVSE Level III quick charging. Paddle was inserted into EVs which was primary coil and secondary coil received power to charge EVs battery. University of Georgia came up with new ICC scheme for EVs charging with range of 200–400 V batteries at operating frequency 77 kHz for 6.6 kW AC EVSE Level II. In RICC, power flow source to battery like this way, first, AC mains voltage to be converted to high-frequency AC voltage, then HF AC voltage is sent to primary side which was connected with compensation network and then power is transferred to secondary side by air gap which was provided with additional compensation network. That is, received AC power at secondary side gets converted to DC and stored at battery management system (BMS).

Capacitive contactless charging is used primarily in low power application such as cellular phone charger but currently researchers are interested in this technique in EVs for primary to secondary power transfer. CCC consists of two metallic sheets which was insulated, one plate serves as primary and second as secondary. Mains is connected to converter via PFC circuit and it produces HF AC and the power is transferred to secondary side via capacitor plates. It works on principle of electro-magnetic induction to charge batter of EVs via magnetic phenomenon. In this circuit, it has two coils for power transfer.

Mains connected with converter and current pass to primary coil and then it induces a magnetic field that induced magnetic field transfer to secondary side to charge battery of EVs. Through its magnetic induction technique power transfer from primary side is fixed and secondary side moving. Because of magnetic coupling effect air gap is high. So, efficiency is low.

7.2 Advantages of Wireless EV Charging

Ease of Use: Eliminates the requirement for plugging in cables, enhancing user satisfaction.

Longevity: Reduces deterioration of connectors.

Smart Technology: Enables self-driving cars that can park and charge themselves.

Wireless charging can happen either while the vehicle is still (static) or while it's in motion (dynamic), although dynamic wireless charging is still in the testing phase.

7.3 Current Global Developments

Several automotive giants and technology firms are piloting wireless charging:

- **BMW 530e iPerformance** offers optional wireless charging systems.
- **WiTricity** and **Qualcomm Halo** are leading in wireless charging standardization and technology development.
- Pilot projects for dynamic wireless charging roads are underway in countries like Sweden and the United States.

7.4 Wireless Charging in India

In India, wireless EV charging is still at an early stage, but pilot programs have begun:

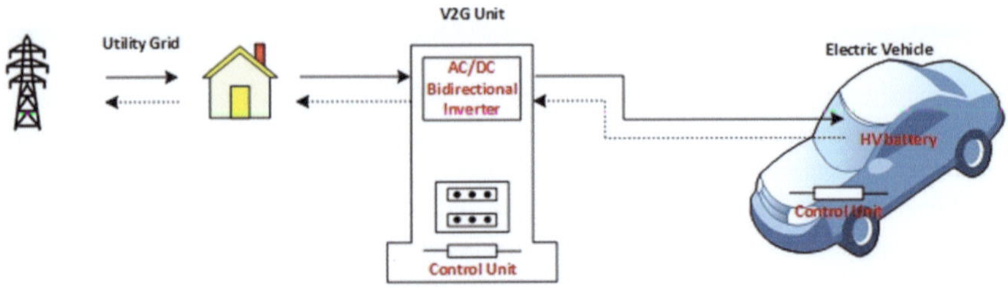

Fig. 5 Block diagram of vehicle-to-grid technology

- IIT Madras is developing low-cost wireless charging solutions tailored for Indian two-wheelers and three-wheelers.
- Private startups are experimenting with wireless EV charging modules for premium electric scooters and corporate fleet vehicles.

Challenges such as cost, efficiency losses, and regulatory standards must be addressed before mass adoption, but research and innovation are accelerating.

8 Vehicle-To-Grid (V2G) Technology: EVs as Energy Assets

8.1 Concept of Vehicle-To-Grid Technology

Vehicle-to-Grid (V2G) technology facilitates the two-way transfer of energy between an electric vehicle (EV) and the electrical grid. Besides drawing energy for recharging, EVs have the ability to send excess electricity back to the grid during high-demand times, essentially functioning as portable energy storage systems.

A bidirectional AC/DC inverter and a control unit that regulates energy flow between the EV's high-voltage (HV) battery and the power grid make up the system architecture, as illustrated in Fig. 5. Energy stored in the EV battery can be released to support the grid during times of high demand, and the battery can be effectively recharged during off-peak hours.

The most popular plug types are Type 1 (J1772), which is mostly utilized in North America and Japan, and Type 2 (Mennekes), which is extensively used in Europe, as shown in Fig. 6.

8.2 Benefits of V2G

Vehicle-to-Grid (V2G) technology provides a revolutionary approach to managing energy since EVs can not only consume energy but also feedback energy to the grid if required. This bidirectional exchange of energy supports grid stabilization in

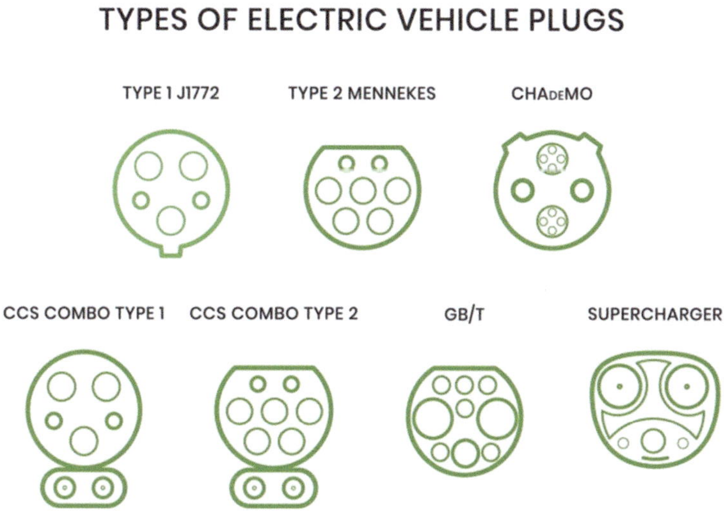

Fig. 6 Types of EV plugs

terms of supplying crucial services like frequency regulation, voltage support, and peak capacity mitigation. When the share of intermittent renewable energy sources, such as wind and solar power grows, V2G can act as a decentralized storage means, smoothing out fluctuations and providing a stable power supply. For EV owners, participation in V2G schemes can create new revenue streams since they can earn money for the power they feed back into the grid or for the services provided. The successful implementation of advanced bidirectional wireless charging systems can significantly contribute to a more resilient and sustainable energy future, facilitating the seamless integration of renewable energy into the grid [24, 25]. Moreover, a widespread implementation of V2G can reduce reliance on fossil-fuel-based peaking plants, reduce overall carbon emissions, and increase grid resilience in times of crisis, and increase infrastructure efficiency by reducing the demand for additional stationary energy storage solutions. Ultimately, V2G increases the integration between the transportation and energy industries, leading to a more intelligent and sustainable future.

Numerous advantages offered by vehicle-to-grid (V2G) technology improve the performance of the power grid and electric vehicle (EV) utilities. V2G facilitates demand balancing, voltage control, and frequency regulation, all of which contribute to grid stability, as shown in Table 4. Additionally, it is essential for the integration of renewable energy since it enables the storage of energy from sporadic sources, such as wind and solar, in EV batteries, which can then be returned to the grid when required.

Table4 Benefits of vehicle-to-grid (V2G) technology

Benefit	Description
Grid stability	Supports frequency regulation, voltage control, and demand balancing
Renewable integration	Helps manage variability from solar and wind generation by storing and supplying energy
Financial incentives	Provides payments or reduced energy costs for EV owners participating in V2G programs
Reduced carbon emissions	Decreases reliance on fossil-fuel-based power generation during peak hours
Infrastructure savings	Reduces the need for investing in additional energy storage or peaking power plants
Emergency backup	Provides emergency power supply during grid outages or natural disasters
Enhanced grid resilience	Distributes energy storage across thousands of vehicles, making the grid more robust

8.3 Global Status of V2G

Nations such as Japan, the Netherlands, the UK, and the US are at the forefront of V2G pilot projects:

1. Nissan LEAF cars in Japan are enabled with V2G capabilities through the CHAdeMO protocol.
2. In California, V2G programs are run by Pacific Gas and Electric Company (PG&E) that are associated with solar energy installations.

Commercial fleets, such as electric school buses and delivery trucks, are seen as the best early adopters of V2G because they have regular schedules.

8.4 V2G Potential in India

India's energy requirements are rapidly increasing, coupled with its vision of achieving net-zero emissions by 2070, which makes it a prime candidate to adopt Vehicle-to-Grid (V2G) technology. The country's transport industry is well on its way to becoming electrified, with the help of government initiatives such as FAME-II (Faster Adoption and Manufacturing of Hybrid and Electric Vehicles) and state-specific EV policies. As electric vehicle adoption increases—with projections of over 30% of vehicle sales by 2030—India may have millions of electric vehicles, two-wheelers, and buses as mobile energy storage devices.

The addition of V2G technology in India can address a number of problems. The power grid with high penetration of solar and wind energy poses challenges like intermittency and peak load pressure. EVs connected through V2G can contribute services such as frequency regulation, peak shaving, and smoothing of renewables with a decentralized and flexible method of grid balancing. Also, in rural and semi-urban areas where grid infrastructure is limited, V2G-enabled EVs might serve as micro-grid stabilizers, thereby strengthening rural electrification.

Financially, V2G can create new revenue streams for EV owners, improve the economic competitiveness of electric mobility, and reduce overall cost of ownership. For the utilities, it offers an economically rational alternative to investing in more power generation and storage facilities. Putting policies in place that promote V2G participation, advance charging infrastructure, and support smart grid technologies will be key to unleashing its full potential.

In India's context, public electric fleets such as electric buses, three-wheelers, and delivery vans are a highly promising first market for V2G because they have large battery packs, regular operating schedules, and centralized charging facilities.

The graph shown in Fig. 7 "EV Growth Projections in India (2020–2030)" shows the expected increase in the number of electric vehicles (EVs) in India over a period of 10 years. Beginning with around 1 million EVs in 2020, the figure gradually rises to about 4 million by 2024. The increase then becomes much faster from there, reaching 10 million by 2026, 20 million by 2028, and reaching a maximum of 30 million EVs by 2030. All of this is in line with the nation's drive toward greener transport based on policy incentives, reducing battery prices, growing infrastructure, and augmenting environmental consciousness. The statistics imply a high rate of adoption, particularly in the second half of the decade, and point toward a significant shift in India's mobility industry.

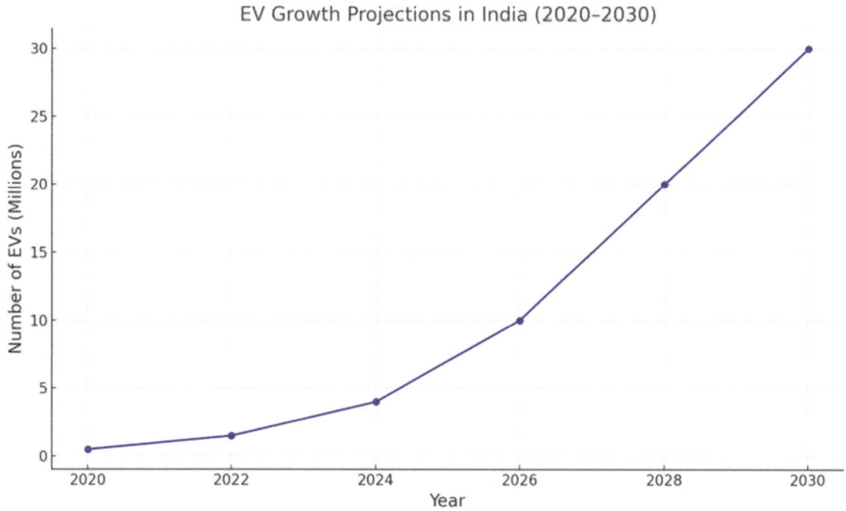

Fig. 7 Projected growth of electric vehicles (EVs) in India from 2020 to 2030

8.5 Challenges and Future Directions

In spite of great progress, many hurdles remain for these technologies to come to widespread deployment:

1. Standardization: Harmonizing charging specifications among firms is critical, particularly for V2G compatibility.
2. Concerns over Battery Degradation: Bidirectional cycling on a regular basis in V2G can impact battery life unless properly controlled.
3. Investment in Infrastructure: Fast charging, wireless grids, and V2G-capable infrastructure require substantial investment.
4. Policy and Regulation: Clear regulations on energy tariffs, grid access, and safety standards need to evolve in parallel with the technology.
5. Consumer Awareness: It will be crucial to educate EV owners on new features such as V2G and wireless charging.

However, the way forward is clear: rapid innovation, falling prices, and favorable policies will drive the next phase of EV technology advancement.

8.6 Conclusion

Technological developments in fast-charging networks, wireless charging systems, and vehicle-to-grid are in essence revolutionizing the electric vehicle market. These technologies as a whole address the primary barriers that have so far constrained EV uptake, including fears about range, inconvenience in charging, and difficulties in integrating with the energy grid. India, while the building blocks are in progress, is set to witness massive growth. With sustained investments, supportive government policy, and cooperation between automobile makers, energy suppliers, and information technology companies, India can make progress toward an era where electric vehicles are not just means of transport but integral components of an intelligent, efficient energy system. The continued convergence of transportation and energy infrastructure will ultimately make electric vehicles more accessible, convenient, and desirable, setting the stage for a greener and smarter future.

References

1. United States Environ. Protection Agency, Sources of Greenhouse Gas Emissions, Jun. 2021. https://www.epa.gov/gemstones/sourcesgreenhouse-gas-emissions
2. SU Khan KK Mehmood ZM Haider SBA Bukhari S-J Lee MK Rafique 2018 Energy management scheme for an EV smart charger V2G/G2 V application with an EV power allocation technique and voltage regulation Appl Sci 8 4 648

3. Itani K, Bernardinis AD, Khatir Z, Jammal A, Oueidat M (2016) Regenerative braking modeling, control, and simulation of a hybrid energy storage system for an electric vehicle in extreme conditions. IEEE Trans Trans Electrification 2
4. Mishima T, Akamatsu K, Nakaoka M (2013) A high frequency-link secondary-side phase-shifted full-range soft-switching PWM DC–DC converter with ZCS active rectifier for EV battery chargers. IEEE Trans on power Electron 28
5. Deng J, Li S, Hu S, Mi CC, Ma R (2014) Design methodology of LLC resonant converters for electric vehicle battery chargers. IEEE Trans vehicular Technol 63
6. Pinto JG, VítorMonteiro H, Gonçalves JL (2014) Afonso, onboard reconfigurable battery charger for electric vehicles with traction-to-auxiliary mode. IEEE trans. Vehicular Technol 63
7. Prasanna UR, Singh AK, Raja shekara K () Novel bidirectional single-phase single stage isolated AC-DC converter with PFC for charging of electric vehicles. IEEE Trans Transp Electrification
8. Gandhodi HV, Konkala KTR, Bandameeda RK, Chaviti J, Yenna PK, Subburaj V (2020) A new single input-multiple output switching converter using circular capacitors for e-vehicle applications. In: TENSYMP 2020–2020 IEEE Region 10 Conference, IEEE, 2020, pp 1696–1699
9. F Musavi M Craciun DS Gautam W Eberle WG Dunford 2013 An LLC resonant DC-DC Converter for wide output voltage range battery charging applications IEEE Trans Power Electron 28 5437 5445
10. S Kundu AK Giri S Kadiyan 2025 An adaptive mixed-step size normalized least means fourth control approach for stand-alone power generation system considering dynamic conditions IEEE J Emerg Sel Top Power Electron 13 2 2422 2429 https://doi.org/10.1109/JESTPE.2024.3516878
11. Lu L, Han X, Li J, Hua J, Ouyang M (2013) A review on the key issues for lithium-ion battery management in electric vehicles. J Power Sources 226:272–288
12. Ministry of Heavy Industries and Public Enterprises, Publication of notification in Gazette of India (Extraordinary) regarding Phase-II of FAME India Scheme (The Gazette of India. New Delhi, 8 Mar 2019)
13. IEA (2019) Hybrid and electric vehicles: the electric drive hauls (technology collaboration programme on hybrid and electric vehicles
14. Jafri NH, Gupta S (2016) An overview of FCs application in transportation. In: IEEE transportation electrification conference and Expo (Asia-Pacific (ITEC Asia-Pacific), Busan), pp 129–133
15. Wikipedia. The Free Encyclopaedia, List of fuel cell vehicles. Accessed 24 Mar 2019
16. Wikipedia. The Free Encyclopaedia, Fuel Cell Vehicle. Accessed 24 Mar 2019
17. Kundu S, Giri AK, Kadiyan S, Singh S (2025) Power management and adaptive control method for a standalone distributed generation system under fault, DC-offset, and dynamic environments. Int J Ambient Energy 46(1). https://doi.org/10.1080/01430750.2025.2452245
18. U Eberle B Müller R Helmolt Von 2012 Fuel cell electric vehicles and hydrogen infrastructure: status 2012 Energy Environ Sci 5 10 8780 8798
19. Plug-ins progress. In Green Car Congress, 29 Sept 2006. Accessed 30 Dec 2017
20. SS Williamson AK Rathore F Musavi 2015 Industrial electronics for electric transportation: current state of- the-art and future challenges IEEE Trans Ind Electron 62 3021 3032
21. EA Grunditz T Thiringer 2016 Performance analysis of current BEVs based on a comprehensive review of specifications IEEE Trans. Transp. Electr. 2 270 289
22. H Shareef MM Islam A Mohamed 2016 A review of the stage-of-the-art charging technologies, placement methodologies, and impacts of electric vehicles Renew Sustain Energy Rev 64 403 420
23. S Kundu M Singh AK Giri 2024 Robust and fast control approach for islanded microgrid system and EV charging station applications Electr Eng https://doi.org/10.1007/s00202-024-02291-1
24. S Meraj S Mekhilef M BintiMubin H Ramiah M Seyedmahmoudian A Stojcevski 2025 Bidirectional wireless charging system for electric vehicles: a review of power converters and

control techniques in V2G application IEEE Access 13 75242 75260 https://doi.org/10.1109/ACCESS.2025.3561396
25. Kundu S, Singh M, Giri AK (2024) Synchronization and control of WECS-SPV-BSS-based distributed generation system using ICCF-PLL control approach. Electr Power Syst Res 226:109919. ISSN 0378-7796. https://doi.org/10.1016/j.epsr.2023.109919

Chapter 8
Transitioning to Electric Mobility: Challenges, Opportunities, and Future Directions

Sunil Kadiyan, Sombir Kundu, Jitender Kumar, and Aman Aggarwal

Abstract The automobile sector is shifting toward Electric Vehicles (EVs) from conventional internal combustion engine (ICE) vehicles. The negative effect of ICE vehicle emissions has a huge negative impact on the environment and causes a lot of health-related issues in society. One of the most dangerous effects of gases produced, such as CO_2 and nitrogen, on the environment is the greenhouse effect. The temperature of the Earth increases every year and threatens society, decreasing the chances of living in the future. EVs are found to be the best alternative to these vehicles due to their smart and robust technology. These vehicles reduced the dependence on limited resources like fossil fuels. These EVs run on chargeable batteries, which produce nearly zero emissions of harmful gases and other by-products. The development of EVs is still in its starting phases. But the future trends show a bright future for transportation, humanity, and a sustainable environment. This chapter includes the future possibilities of EVs. The difficulties associated with this technology, charging capacity, charging infrastructural problems, and future possibilities. This chapter is important to understand the limitations and future aspects of EVs.

Keywords EVs · Infrastructure · Environmental sustainability · Power electronics · Smart Charging

S. Kadiyan (✉) · J. Kumar · A. Aggarwal
Department of Mechanical Engineering, Ganga Institute of Technology and Management, Jhajjar, India
e-mail: sunkadiyan@gmail.com; sunil.me@gangainstitute.com

S. Kundu (✉)
Electrical Engineering Department, Maharishi Markandeshwar Engineering College, Maharishi Markandeshwar (Deemed to Be University), Ambala, Haryana, India
e-mail: sskundu46@gmail.com; dr.sombir@mmumullana.org

S. Kundu et al. (eds.), *Hybrid Renewable Power Infrastructure for Sustainable Electric Vehicle Development*, Studies in Infrastructure and Control,
https://doi.org/10.1007/978-981-95-1323-9_8

1 Introduction

Cars and other vehicles mainly run on petrol and diesel. Many countries in the world are planning to ban these vehicles in the future due to the concept of environmental sustainability [1]. So, the use of EVs increases due to very less use of fossil fuel and low harmful gas emissions. The primary focus for using EVs is creating non-pollution-producing vehicles, and the second is increasing the power storage capacity of these types of vehicles [2]. The present technology has some drawbacks such as low charging capacity and insufficient charging stations [3]. These problems create a question mark in the shift toward EVs as compared to conventional ICE vehicles. A lot of researchers have shown the method of increasing the efficiency of the current EV battery system [4]. Many authors represent the models that produce effective results even in different load conditions [5]. B. Latha et al. represent a classification of EV charging methods based on the mode of power transmission. It divides charging into two main types: plug-in charging and wireless charging. Plug-in charging is further categorized by the type of power supply—AC (Alternating Current) or DC (Direct Current)—and by the location of the charger, either on-board or off-board. It also includes charging time classifications: Mode 1 (slow), Mode 2 (medium/partial fast), and Mode 3 (fast). Each class under AC and DC charging represents a different power level, ranging from low (Class 1) to high (Class 3). On the other hand, wireless charging is categorized as radiative or non-radiative, with non-radiative charging also classified into Modes 1 to 4 based on charging speed [6]. S. Kundu et al. represent various controls that promote the use of EVs in the future. A hybrid DG system that integrates wind, solar, and battery storage sources to supply power to unbalanced loads. The wind energy is harnessed through an SEIG, which is connected to the system via a CPI. Simultaneously, solar power is generated by a photovoltaic (PV) array, regulated through a Maximum Power Point Optimization (MP&O) unit, and delivered via a Bidirectional Converter (BDC). A Battery Storage System (BSS) is included to store excess energy and support the load during low-generation periods. The DC power from both PV and BSS is converted to AC using a Voltage Source Converter (VSC), which is controlled by a dedicated algorithm. A ripple filter is placed at the output of the VSC to reduce electrical noise and deliver smooth power. The system effectively maintains stable voltage and current even under unbalanced load conditions [5].

A huge literature is available to showcase the bright future probabilities for the development of EVs [7]. F. Noor et al. present a very basic and important paper for understanding the concept of EVs, the basic parts associated with EVs, and the power required to run with economic and environmentally friendly factors. The authors suggested most of the challenges associated with the EV and the future scopes [8]. Y. Chen et al. present the EV energy consumption models, emphasizing their role in supporting low-carbon transportation. It categorizes influencing factors into four groups—vehicle components, dynamics, traffic, and environment—and classifies models by scale and methodology. The authors highlight growing trends in data-driven models [9]. As EV technology continues to evolve, the demand for advanced charging systems and better infrastructure will only increase [10]. Consumers are looking for faster, wireless charging options, real-time monitoring, and high levels of convenience and reliability. To meet these expectations, continuous innovation in EV charging technology is necessary [11].

1.1 Positive Aspects of Using EVs

EVs produce clean energy as compared to conventional vehicles [12]. All types of EVs produce fewer harmful by-products than combustion engine vehicles. In IC engine vehicles, about 75% of the energy is lost due to heat and other inefficiencies, while in EVs, only around 25% of energy is lost [13]. Most of the energy in EVs is efficiently converted into electric power, which directly influences energy consumption per mile. As a result, the operating cost of EVs is much lower than that of IC engine vehicles. More than this the number of moving parts in the EVs is very low and the number of moving parts is the reason for the low maintenance cost. The production of energy by using renewable sources makes this technique unique and best [4, 14, 15].

1.2 Environmental Aspects of EVs

One of the main reasons for the world's biggest problem of pollution is the after-burning waste of fossil fuels [13]. Petrol and diesel vehicles produce harmful gases as waste. In contrast, EVs produce very low emissions throughout their lifespan. Moreover, electricity for EVs can be generated from renewable sources like solar energy [16]. These environmental benefits give EVs a significant edge over conventional vehicles and promote their adoption.

1.3 Need for Evolving EV Charging Requirements

Although EVs offer many advantages over IC engine vehicles, improvements are still needed in their design and charging capabilities. This chapter focuses on the current technologies used in EVs and explores areas that can be developed further in the future to improve EV performance and convenience.

2 Present Technology in EV Charging

In present technology, the battery of the EVs is charged just like your mobile phone [17]. The battery may be slowly charged, fast-charged, or rapidly charged [10]. The type of charging depends on the resources. AC can charge the battery at a slow speed. The battery is just put in the socket, just like a mobile phone, and it will take 7 to 12 h of charging, or you can say overnight charging is required. This is slow charging or AC Charging. The fast charging uses a DC supply for the fast charging of the battery. It will charge the battery from 70 to 80% in just 30 min. This required public charging stations, EV service stations, etc. It's a quick charging top-up system for fast charging. Rapid fast charging is another technique for charging batteries in EVs. It uses high power to recharge the battery very quickly. It can charge the batteries in just 10 to 15 min. This type of charging helps in long-distance traveling. These types of charging stations are still in a developing condition [6]. Wireless charging is another method of charging EV batteries [18]. These are also known as inductive charging. These chargers are slow like slow chargers via AC. In this type of charging, there is no need to connect a wire; instead, the battery is charged by using induction. One method is battery swapping, in which the drained battery is replaced with the charged battery. Racha et.al present their work on the importance of green hydrogen for producing electricity just by interacting with oxygen present in the environment. This paper explores the technique for producing renewable energy as fuel for running the EVS. Green additives may be helpful compared to fossil fuels. This paper also concludes the importance of sustainable fuels. The authors represent the difference between conventional IC engine vehicles, their emissions, and the side effects on the environment, with the positive aspects of renewable energy resources. V.J. Reddy et al. present their work with green hydrogen, methanol, and ammonia as a fuel for a better future or a secure future from pollution produced due to other fuels like petrol and diesel. He presented his work on the zero emission of pollution up to 2020 by using these as fuel. He also suggested the weakness of the present work as the high cost and the very complex technology. However, it can be achieved by a collaborative effort of the global engineer's efforts [19].

Nammour et al. present their work on how hydrogen can be used to react with CO_2 to produce fuels like methanol, which can help in running EVs with very low emissions of gases that produce a greenhouse effect. The authors not only suggest producing clean fuel but also suggest working on strengthening the power grid system for better efficiency of the EVs [20].

S. Kundu et al. present the concept of a microgrid. He also suggests power quality and load balancing through adaptive control. Future charging infrastructure highly depends on the renewable-powered microgrids [21].

Figure 1 shows the control topology of an isolated microgrid using wind, solar, and battery storage with smart control for voltage and frequency regulation. The voltage source inverter (VSI) and bidirectional converters ensure a reliable power supply and quality under varying load conditions. This system is crucial for future EV

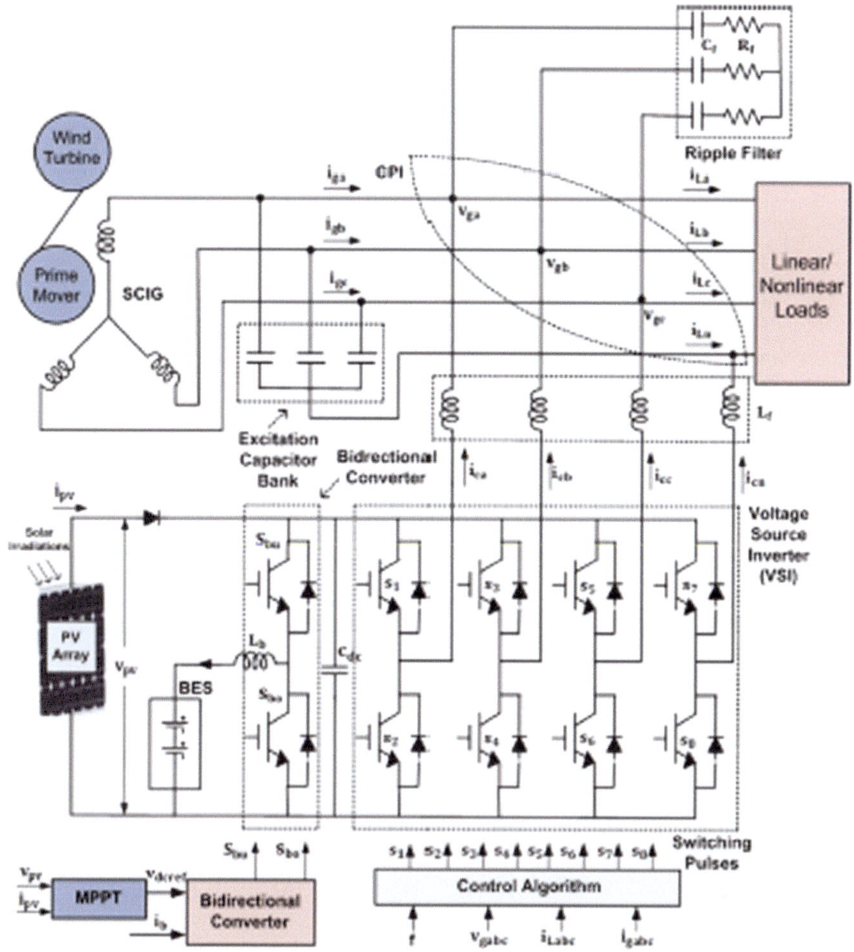

Fig. 1 Control topology of an isolated microgrid

charging, especially in remote or off-grid areas, enabling clean, stable, and decentralized energy. As EV adoption grows, such microgrids will support sustainable, flexible, and resilient charging infrastructure. S. Kundu et al. represent the renewable energy generation system.

Figure 2 shows the dynamic response of a wind and solar PV (SPV) based microgrid during an unbalanced load condition. It illustrates how key parameters like voltage, current, frequency, and DC-link voltage remain stable even after the removal of one phase load. These techniques will be best for the unbalanced charging demands.

The authors presented another useful work on a standalone power generation system. It helps EVs in nonlinear load conditions [21]. S. Kundu et al. represent

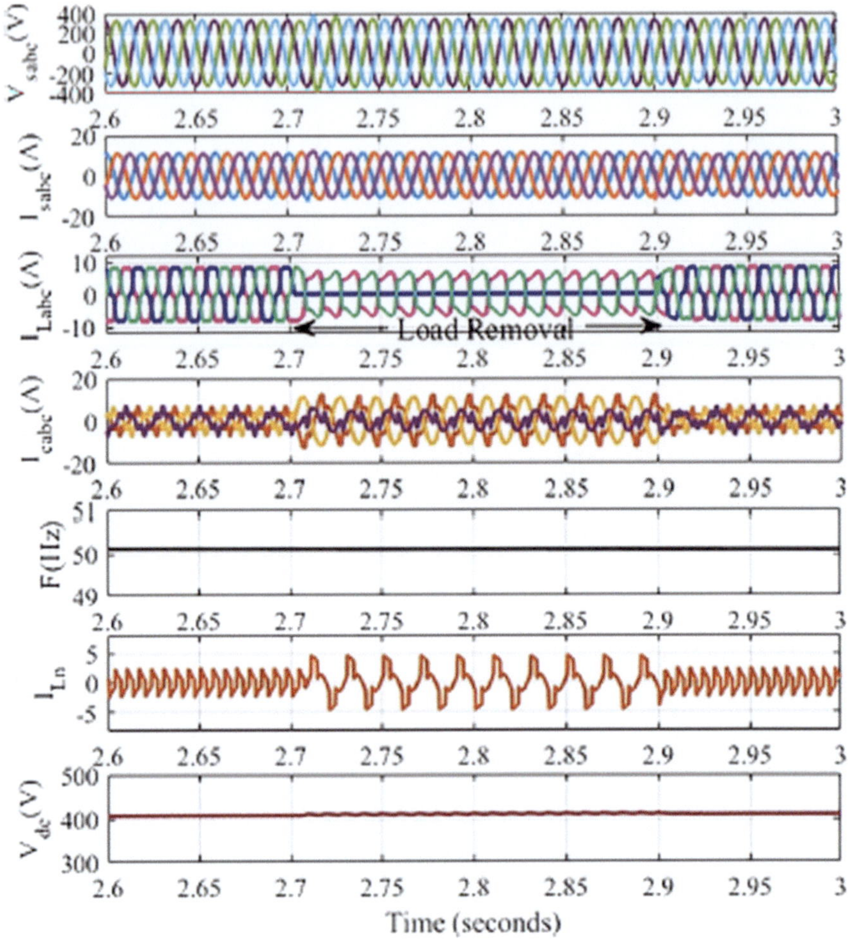

Fig. 2 Dynamic response of a wind and solar PV (SPV) based microgrid during an unbalanced load

in Fig. 3 that the voltage and frequency regulation while managing power quality issues from nonlinear loads. The bidirectional converter optimizes solar power use and battery performance. Overall, the system maintains reliable operation under varying and unbalanced load conditions [22].

Figure 4 shows the topology of an isolated microgrid that combines wind, solar, and battery storage to supply power to nonlinear loads. A wind turbine drives a self-excited induction generator (SEIG) to produce AC power, while a solar panel (SPV) generates DC power, optimized using a modified Perturb & Observe (P&O) method. A bidirectional converter (BDC) manages the battery's charging and discharging to maintain power balance. The voltage source converter (VSC), controlled by a smart algorithm, ensures voltage and frequency stability and removes harmonics. A ripple

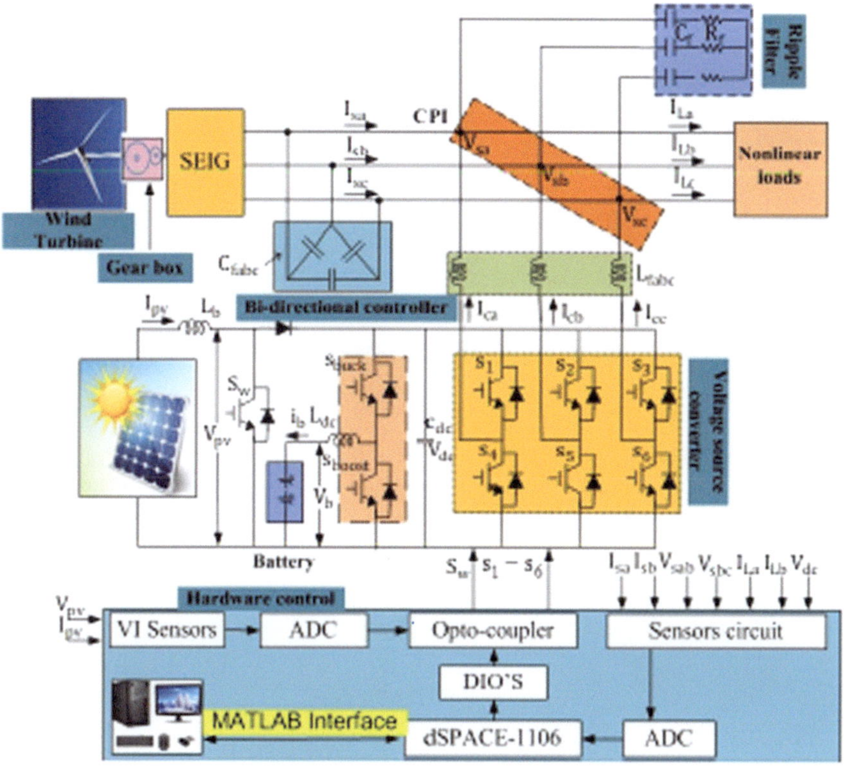

Fig. 3 Standalone power-generating system

filter smooths the power before it reaches the nonlinear load, making the system reliable for off-grid applications and future EV charging needs. A. Zentani et al. represent their work on the fast charging technology. Figure 5 shows the charging techniques.

This paper reviews current EV fast charging technologies, including ultra-fast, DC, wireless charging, and battery swapping. It explains the infrastructure, charger types, control strategies, and the role of smart grids and bidirectional charging. Key challenges like standardization, infrastructure limits, and battery tech needs are discussed. The study highlights the importance of global cooperation and future research for faster, safer, and more efficient EV charging [23].

R. Padilla et al. show how EV charging stations are now managed more independently, aiming to improve energy distribution and efficiency. With this shift, managing charging operations has become more complex, requiring smarter, more sustainable decision-making models [24]. Figure 6 shows the charging levels.

The study reviews current and future technologies, protocols, and case studies related to charging stations. It highlights trends like intelligent energy systems, IoT integration, and sustainable energy management for advancing electromobility [25].

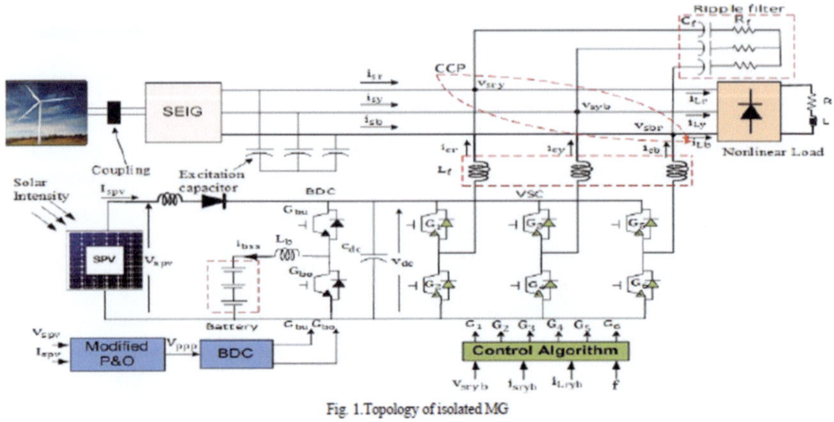

Fig. 4 Topology of isolated MG

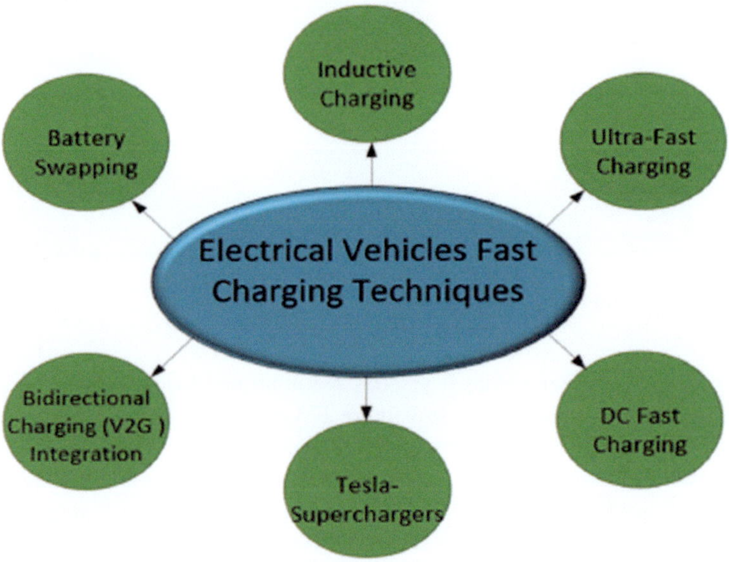

Fig. 5 EV charging techniques

(a) (b) (c)

Fig. 6 Various charging levels: **a** Level 1 and Level 2 Residential Charging, **b** Work and public place charging, **c** DC Fast charging

S. Acharige et al highlight the rapid growth of EVs and their key role in the global energy transition. It discusses the challenges EVs pose to the power grid, such as planning, stability, and safety, and emphasizes the need for advanced charging systems and equipment. The study reviews current EV charging technologies, standards, station architectures, and converter setups, including systems powered by renewable energy. It also outlines future trends and challenges, aiming to improve charging efficiency and better integrate EVs with the grid [26].

2.1 Limitations of Existing Infrastructure

The electric vehicle (EV) technology is growing steadily, but there are still some problems that need to be solved before more people can start using EVs. One big issue is the high cost. Hydrogen fuel cell technology is expensive because producing hydrogen and maintaining hydrogen plants costs a lot. Also, the infrastructure for this technology is not fully developed yet. There are very few fast charging stations, and slow charging stations take more time, which makes it inconvenient for users. So, the main challenges for EV adoption are the high cost, lack of proper charging infrastructure, and long charging time.

3 Future Trends in EV Charging

EVs with fuel cell technology are one of the best-suited technologies for a sustainable environment. The time is for the technology that causes the minimum effect on the environment and produces nearly zero emissions. The green hydrogen can be produced by the method of electrolysis powered by renewable energy sources like wind and solar. This approach not only improves the technology but also reduces the emission of hazardous gases like carbon and nitrogen, and pushes nature toward a green environment. Many optimization techniques also work on the concept of reducing the cost and improving the efficiency of EVs. T. Long et al. represent their work on the optimization. He proposed that a hydrogen-EV integrated architecture effectively reduces operating costs and improves charging efficiency. Case studies in Shanghai demonstrate at least a 13% performance improvement over existing methods. This highlights the strong economic potential and practical value of the joint optimization approach [27]. D. Das represents a dynamic concept of the integration of hydrogen in remote areas. This study models a hybrid renewable energy system combining solar, wind, and green hydrogen to meet home and EV loads in remote areas like Siliguri, India. Using HOMER Pro, various grid-connected and off-grid scenarios are analyzed to optimize cost, capacity, and emissions. The results highlight green hydrogen's role in enhancing system efficiency and supporting sustainable energy transitions. Many other technologies in hydrogen production may also be adopted in the future, like solid oxide electrolyzers (SOECs). This technology can operate at high temperatures. This technique may produce high performance when it works collaboratively with industrial waste heat. Photocatalysis is another emerging technique that can be used. [28, 29, 30, 31] S Kundu et al. highlight the importance of clean power supplied to the EV for better performance. The quality of input highly determines the output in the form of emissions from the EVs. The proposed control system ensures stable, high-quality power delivery for electric vehicle (EV) charging in remote areas using a renewable-based islanded microgrid. By integrating solar PV, wind energy, and battery storage, the system supports reliable EV charging even during power fluctuations. Advanced SGF-PLL control enhances power quality and maintains system stability under dynamic conditions. This architecture enables cost-effective, resilient, and sustainable EV charging infrastructure in off-grid and variable supply environments. A similar type of clean power concept was highly associated with the many publications of S Kundu. In another publication, he proposed an SPV-wind-BES-based islanded electrical supply system, equipped with a 3Ø-APF-PLL control approach, which ensures stable voltage, frequency regulation, and harmonic reduction under varying renewable energy and load conditions. By integrating solar, wind, and battery storage, the system provides a continuous and high-quality power supply, making it highly suitable for remote electric vehicle (EV) charging infrastructure. The fast dynamic response and power compensation capabilities of the system support reliable EV charging even in off-grid or power-deficient regions, offering a sustainable and resilient energy solution[4]. This highlights the vast potential for expanding EV charging capacity and exploring new charging station technologies, which drives further development and innovation in electric vehicles.

4 Conclusion

1. Efficient charging technologies are essential for electric vehicle (EV) adoption. Fast chargers reduce wait times and improve convenience. Different charger types serve various user needs. Innovation will make EVs easier to use.
2. Using renewable energy for EV charging lowers emissions. Green hydrogen offers clean fuel options. Combining solar, wind, and hydrogen improves sustainability. These methods support clean, reliable power. They help meet climate goals.
3. Smart charging uses AI and IoT to optimize energy use. Wireless charging adds convenience by removing cables. These technologies improve the user experience. They make charging more efficient and safer.
4. Digital tools manage EV charging networks and grid demand. Dynamic pricing helps avoid overloads. Vehicle-to-grid systems let EVs supply power back to the grid. This supports grid stability and renewable energy use.
5. Innovation in EV charging will help reduce carbon emissions. Advances in power electronics and smart controls are key. Cooperation among stakeholders speeds progress. Future tech will enable a cleaner transport system.

References

1. Verma S et al (2021)A comprehensive review on energy storage in hybrid electric vehicle. J Traffic Transp Eng 8(5):621–637. https://doi.org/10.1016/j.jtte.2021.09.001
2. Waseem M, Amir M, Lakshmi GS, Harivardhagini S, Ahmad M (2023) Fuel cell-based hybrid electric vehicles: an integrated review of current status, key challenges, recommended policies, and future prospects. Green Energy Intell. Transp. 2(6):100121. https://doi.org/10.1016/J.GEITS.2023.100121
3. Suresh M, Meenakumar DR, Raghavendran DPS, Gowrishankar V (2018) Design and development of automatic whiteboard cleaner for effective cleaning mechanism using Arduino. Ijireeice 6(9):30–33. https://doi.org/10.17148/ijireeice.2018.696
4. Kundu S, Singh M, Giri AK, Kadiyan S, Chittora P, Shailly D (2024) Robust and fast control approach for islanded microgrid system and EV charging station applications. Electr Eng. https://doi.org/10.1007/s00202-024-02291-1
5. Kundu S, Giri AK, Singh M, Kadiyan S (2024) Adaptive control strategy for isolated renewable energy-based generation system with intermittency. Electr Eng. https://doi.org/10.1007/s00202-024-02454-0
6. Latha B, Mujahid M, Flah A, Blazek V, Prokop L, Rangarajan SS (2024) E-Prime—Advances in electrical engineering , electronics and energy advances in EV wireless charging technology—a systematic review and future trends. e-Prime–Adv Electr Eng Electron Energy 10:100765. https://doi.org/10.1016/j.prime.2024.100765
7. Kundu S, Giri AK, Kadiyan S, Singh S (2025) Power management and adaptive control method for a standalone distributed generation system under fault, DC-offset, and dynamic environments generation system under fault, DC-offset, and dynamic environments. Int J Ambient Energy ISSN. https://doi.org/10.1080/01430750.2025.2452245
8. Un-noor F, Padmanaban S, Mihet-popa L, Mollah MN, Hossain E (2006) Future direction of development. https://doi.org/10.3390/en10081217
9. Chen Y et al (2023) RT ECAP and rolling bestow high strength and good ductility on a low lithium aluminum alloy. J Mater Res Technol 25:5561–5574. https://doi.org/10.1016/j.jmrt.2023.07.009
10. Gnanavendan S et al (2024) Challenges, solutions and future trends in EV-technology: a review. IEEE Access 12:17242–17260. https://doi.org/10.1109/ACCESS.2024.3353378
11. Kumar G, Kumar M (2023) A critical review on recent trends in friction stir welding and processing of aluminum metal matrix composite and aluminum International Journal of Research in Engineering and Innovation REVIEW ARTICLE A critical review on recent trends in friction stir w. https://doi.org/10.36037/IJREI.2023.7203
12. de Souza LLP, Lora EES, Palacio JCE, Rocha MH, Renó MLG, Venturini OJ (2018) Comparative environmental life cycle assessment of conventional vehicles with different fuel options, plug-in hybrid and electric vehicles for a sustainable transportation system in Brazil. J Clean Prod 203:444–468. https://doi.org/10.1016/j.jclepro.2018.08.236
13. Verma S, Dwivedi G, Verma P (2022) Life cycle assessment of electric vehicles in comparison to combustion engine vehicles: A review. Mater. Today Proc. 49:217–222. https://doi.org/10.1016/J.MATPR.2021.01.666
14. Kundu S, Giri AK, Kadiyan S (2024) An adaptive mixed-step size normalized least means fourth control approach for standalone power generation system considering dynamic conditions. IEEE J Emerg Sel Top Power Electron 1. https://doi.org/10.1109/JESTPE.2024.3516878
15. Kundu S, Singh M, Giri AK (1978) Design and control of a standalone wind—solar system with SEIG Feeding linear/nonlinear Loads 317–327
16. Onat NC, Kucukvar M, Tatari O (2015) Conventional, hybrid, plug-in hybrid or electric vehicles? State-based comparative carbon and energy footprint analysis in the United States. Appl Energy 150:36–49. https://doi.org/10.1016/j.apenergy.2015.04.001
17. Bratov V, Borodin EN (2015) Comparison of dislocation density based approaches for prediction of defect structure evolution in aluminium and copper processed by ECAP. Mater Sci Eng A 631:10–17. https://doi.org/10.1016/j.msea.2015.02.019

18. Mohamed AAS, Shaier AA, Abdallah M (2024) Wireless charging for electric vehicles. Power Electron Convert Syst 2nd Ed. 2 Appl 2:321–354. https://doi.org/10.55041/ijsrem34757
19. Reddy VJ, Hariram NP, Maity R, Ghazali MF, Kumarasamy S (2023) Sustainable E-fuels: green hydrogen, methanol and ammonia for carbon-neutral transportation. World Electr Veh J 14(12). https://doi.org/10.3390/wevj14120349
20. Nemmour A, Inayat A, Janajreh I, Ghenai C (2023) Green hydrogen-based E-fuels (E-methane, E-methanol, E-ammonia) to support clean energy transition: a literature review. Int J Hydrogen Energy 48(75):29011–29033. https://doi.org/10.1016/J.IJHYDENE.2023.03.240
21. Sharma S, Kundu S, Singh S, Singh J, Singh S (2025) Harmonics mitigation and control of grid-connected PV system under intermittent conditions. In: Kumar A, Singh SN, Kumar P (eds) Energy management in renewable sources integrated system. ICGEST 2023. Lecture Notes in Networks and Systems, vol 1236. Springer, Singapore. https://doi.org/10.1007/978-981-96-1012-9_26
22. Kundu S, Singh M, Giri AK (2022) Implementation of In-loop filter based DSC-FLL control approach for power quality improvement in isolated microgrid. IEEE 516–521
23. Zentani A, Almaktoof A, Kahn MT (2024) A comprehensive review of developments in electric vehicles fast charging technology. Appl Sci 14(11). https://doi.org/10.3390/app14114728
24. Tejeida-Padilla R, Berdeja-Rocha EM, Badillo-Piña I, Pérez-Matamoros Z, Amador-Santiago JE (2024) A comprehensive review of electric charging stations with a systemic approach. World Electr Veh J 15(12). https://doi.org/10.3390/wevj15120571.
25. Kunicina N et al (2024) Decision-making system for electric vehicle management by integrating smart technologies and local characteristics. Appl Sci 14(23):1–20. https://doi.org/10.3390/app142311150
26. Acharige SSG, Haque ME, Arif MT, Hosseinzadeh N, Hasan KN, Oo AMT (2023) Review of electric vehicle charging technologies, standards, architectures, and converter configurations. IEEE Access 11(February):41218–41255. https://doi.org/10.1109/ACCESS.2023.3267164
27. Long T, Jia Q (2021) Joint optimization for coordinated charging control of commercial electric vehicles under distributed hydrogen energy supply. IEEE Trans Control Syst Technol 1–9
28. Das D, Chakraborty I, Bohre AK, Kumar P (2024) Sustainable integration of green hydrogen in renewable energy systems for residential and EV applications. Int J Energy Res. https://doi.org/10.1155/2024/8258624
29. Kundu S, Singh M, Giri AK (2022) Control algorithm for coordinated operation of wind-solar microgrid standalone generation system. Energy Sources, Part A: Recover, Util, Environ Eff 44(4):10024–10044. https://doi.org/10.1080/15567036.2022.2143943
30. Kundu S, Singh M., Giri AK (2024) Synchronization and control of WECS-SPV-BSS-based distributed generation system using ICCF-PLL control approach. Electr Power Syst Res 226:109919. https://doi.org/10.1016/j.epsr.2023.109919
31. Singh N, Pratap B (2025) Robust observer design for variable speed wind turbine using a two-mass model. In: Proceedings of the international conference on systems, control and automation. ICSCA 2023. Lecture Notes in Electrical Engineering, vol 1260. Springer, Singapore. https://doi.org/10.1007/978-981-97-7384-8_15

Chapter 9
Hybrid EV Charging Systems: A Study Report on Integration of Fuel Cell with EV for Clean Solution

Nitin Sundriyal, Sombir Kundu, Sandeep Gupta, Ankit Bhatt, and Navdeep Singh

Abstract The increasing adoption of electric vehicles (EVs) has brought new challenges for charging infrastructure, particularly in terms of grid reliability, environmental sustainability, and accessibility in remote regions. This study investigates the integration of hydrogen fuel cells into hybrid EV charging systems as a viable approach to address these issues. By converting hydrogen into electricity with water as the only byproduct, fuel cells provide a clean and continuous power source that can supplement or even replace conventional grid-based charging. This property is especially valuable for high-demand sites such as commercial charging stations, as well as for off-grid applications where traditional infrastructure is deficient. The work reviews several hybrid charging architectures that combine fuel cells, batteries, and renewable energy sources, evaluating their working benefits, which include emission reduction, rapid refueling, and combining with solar and wind power. Some Real-world examples are also discussed to demonstrate system performance and practical considerations. The analysis also considers current challenges in hydrogen production, storage, and distribution, and highlights recent technological advancements that aim at overcoming these barriers. By connecting theoretical hypotheses with practical case studies, this work offers guidance for researchers, policymakers, and industry professionals who are working toward advancing sustainable EV charging networks. The chapter investigates the potential of fuel cell integration as a solution for resilient and environmentally friendly hybrid EV charging solutions.

Keywords Clean energy · Fuel Cell Fuel Cell Electric Vehicles (FCEVs) · Hybrid Electric Vehicles (HEVs) · Hydrogen storage · Battery, Sustainable Goals

N. Sundriyal (✉) · S. Kundu · N. Singh
Electrical Engineering Department, MMEC, MM (Deemed to Be University), Ambala, Haryana, India
e-mail: nitinssundriyal@gmail.com

S. Gupta · A. Bhatt
Electrical Engineering Department, Graphic Era (Deemed to Be University), Dehradun, Uttarakhand, India

S. Kundu et al. (eds.), *Hybrid Renewable Power Infrastructure for Sustainable Electric Vehicle Development*, Studies in Infrastructure and Control,
https://doi.org/10.1007/978-981-95-1323-9_9

1 Introduction

1.1 The Role of Hybrid Electric Vehicles in Sustainable Transportation

The transport sector is one of the leading sources of air pollution and greenhouse gas emissions globally, which contributes significantly to increased levels of carbon dioxide, nitrogen oxides, and particulate matter. These emissions are linked to air quality and negative health outcomes, and are one of the major reasons for climate change. The presence of conventional vehicles, which depend on gasoline or diesel-powered internal combustion engines (ICEs), requires urgent attention to alternative mobility solutions that can decrease these environmental impacts [1–3].

The popularity of Hybrid Electric Vehicles (HEVs) is increasing day by day; they serve as one of the most efficient solutions for controlling the problems associated with personal and commercial transportation systems. In contrast to conventional vehicles, HEVs are more fuel efficient and release lower quantities of pollutants because they use both Internal Combustion Engines (ICE) and electric drive systems. The impact provided by this dual power mechanism gives HEVs flexibility in terms of driving experience and range, making them widely accepted by people and organizations seeking to cut down on their carbon emissions without having to compromise on performance or convenience.

An HEV is primarily structured around an Internal Combustion Engine (ICE), one or multiple electric motors, and a battery pack. HEVs are different from fully electric vehicles in that they do not depend solely on out-of-vehicle charging; HEVs internally generate electricity (through regenerative braking or ICE acts as a generator). This is particularly useful in remote regions lacking access to charging stations because these HEVs do not need to depend on power grids. Also, electric motors contribute to high-power-dependent tasks like acceleration and climbing, which further decreases fuel consumption with the use of smaller, more fuel-efficient engines.

Hybrid cars stand out because they can save energy when you brake, sending that power back to the battery instead of wasting it. This not only helps you go farther on less fuel but also means your brakes last longer. There are all sorts of hybrids out there—from mild ones that just give the engine a little boost, to full hybrids that can run on electric power alone for short trips, and plug-in hybrids you can charge at home for even more electric driving. Over time, hybrids like the Toyota Prius have proven they use less fuel and are quieter than regular cars, making them a good step between traditional vehicles and fully electric ones. Thanks to better batteries and smarter electronics, hybrids now come in many shapes and sizes to fit different needs. Even though fully electric cars are growing in popularity, they still face hurdles like charging station shortages and high battery costs, especially outside cities. That's why hybrids, and especially those using hydrogen fuel cells, are seen as a flexible choice for people looking for a cleaner ride without giving up convenience. Concisely, hybrids play an important role as we move toward greener transportation. By combining gas engines with electric power, they help to address many environmental and energy issues in the transport sector. Ongoing progress in both hybrid and fuel cell technology will be very important to building a cleaner and more reliable way for the future.

1.2 Hybrid Electric Vehicle Technology

The use of hybrid electric vehicles (HEVs) is one of the most common use cases of electricity in public transport, considering their practicality and eco-friendly features. Both the combustion engine and the electric motor have their advantages; thus, HEVs reap the benefits of using both, which results in an eco-sensitive model of driving.

The efficiency with which HEVs utilize energy is what sets them apart. Like all other cars, they come equipped with motors, high-voltage batteries, and assembled electronic components that connect all parts. An engine that is able to cruise on highways and charge a battery on the go gives a vehicle a lot of driving tours. Meanwhile, the electric motor takes the lead during low-speed driving, stop-and-go traffic, or whenever a burst of extra power is required—like climbing a steep hill or overtaking another car. The battery stores energy, which it gets not only from the engine but also from regenerative braking. Instead of wasting energy as heat when you slow down, HEVs capture some of that energy and put it back into the battery.

Modern hybrids are smart. They constantly monitor things like how much charge is left in the battery, how hard you're pressing the accelerator, and what kind of driving you're doing—whether it's city traffic or open highways. Based on this information, the car decides when to use the engine, the electric motor, or both. For example, in city driving, you might find the car running mostly on electricity, which keeps things quiet and cuts down on emissions. On longer highway trips, the engine does more of the work, ensuring you don't have to worry about running out of battery.

There are several types of HEVs, each offering a different balance between electric and engine power [4]. Mild hybrids use a small electric motor mainly to support the engine, especially during starting and braking, but can't drive on electricity alone. Full hybrids have larger batteries and motors, letting them run short distances using just electricity—perfect for quiet, emission-free trips around town. Plug-in hybrids go a step further, with batteries you can charge from a wall outlet, allowing for even longer electric-only journeys before the engine needs to kick in.

The technology behind HEVs has come a long way. Advances in battery chemistry—shifting from nickel-metal hydride to modern lithium-ion batteries—have made hybrids lighter, more powerful, and more durable. Improvements in the electronics that manage power flow mean today's hybrids are smoother and more responsive than ever. These developments have real-world impacts: HEVs like the Toyota Prius have shown that it's possible to cut fuel consumption by up to a third compared to similar conventional cars, all while reducing emissions and even making city streets a bit quieter [5]

For many drivers, HEVs offer a sensible bridge between the familiar world of gasoline vehicles and the future of fully electric cars. They're especially valuable in places where charging stations are still few and far between, or where the power grid isn't always reliable. By making the most of both engine and electric power, hybrids help reduce our reliance on fossil fuels and pave the way for cleaner, more sustainable mobility.

In short, hybrid electric vehicles are not just a technical achievement—they're a practical solution, helping drivers make a positive impact on the environment without sacrificing convenience or peace of mind. As battery technology and energy management systems continue to improve, HEVs are set to remain a key part of the journey toward greener transportation.

1.3 Types of Hybridization

HEVs can be classified according to their degree of electrification and system architecture:

- **Mild Hybrids:** Utilize a small electric motor to assist the ICE, primarily for start-stop functionality and regenerative braking, but cannot propel the vehicle on electric power alone.
- **Full Hybrids:** Feature larger batteries and more powerful motors, allowing for short-distance, low-speed electric-only operation. The system can seamlessly switch between or combine ICE and electric motor power.
- **Plug-in Hybrid Electric Vehicles (PHEVs):** Incorporate even larger battery packs that can be charged from external sources, enabling extended electric-only driving ranges and further reducing reliance on the ICE [6].

1.4 Technological Progress and Impact

The evolution of battery technology, particularly the shift from nickel-metal hydride (NiMH) to lithium-ion (Li-ion) chemistries, has significantly enhanced the energy density, cycle life, and overall performance of HEVs [6]. Power electronics advancements have enabled more efficient energy conversion and precise control of propulsion systems. These improvements have translated into tangible benefits: reduced fuel consumption, lower greenhouse gas emissions, and improved drivability.

Commercially, HEVs have gained widespread acceptance, with models such as the Toyota Prius demonstrating the viability and scalability of hybrid technology. According to recent studies, HEVs can achieve fuel savings of 20–35% compared to conventional vehicles, and their adoption has contributed measurably to emission reductions in several markets [7]

HEVs serve as a practical bridge between traditional ICE vehicles and fully electric vehicles (EVs). Their ability to self-charge and operate efficiently in diverse driving conditions makes them particularly valuable in regions where charging infrastructure is limited or grid reliability is a concern. As the automotive sector continues

Fig. 1 HEV schematic

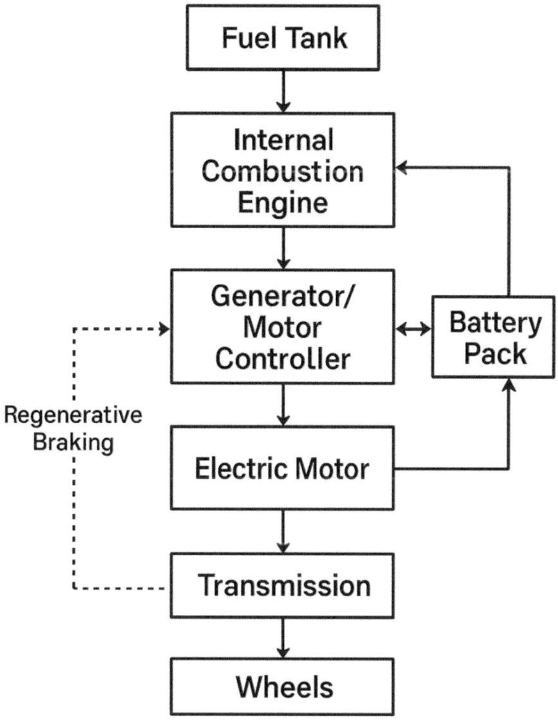

to pursue decarbonization, HEV technology is expected to play an ongoing role in reducing the environmental impact of transportation while facilitating the transition toward electrified mobility. A schematic showing the HEV system is given in Fig. 1 (where the solid line indicates energy/power flow, the double arrow indicates charging/ discharging, and the dotted line represents braking from wheel assembly back to battery).

1.5 Clean Energy Solutions in Transportation

The transportation sector is a major contributor to global greenhouse gas emissions, accounting for a significant portion of carbon dioxide (CO_2), nitrogen oxides (NO_x), and particulate matter released into the atmosphere. These pollutants negatively impact air quality and human health, and contribute to global climate change. Traditional vehicles powered by gasoline or diesel engines are primary sources of these emissions, which has led to a worldwide shift toward cleaner alternatives.

The creation of hybrid electric vehicles (HEVs) marked a significant achievement in reducing fuel consumption and emission rates in everyday commuting. For example, the Toyota Prius model spearheaded this sector, demonstrating improved carbon dioxide emissions and fuel consumption compared to petrol or diesel vehicles. Aside from minimizing greenhouse gas emissions, the implementation of electric motors in HEVs further reduces the noise in urban areas. All HEVs share the characteristics of a bridge technology. Bridge technology means that, unlike other electric vehicles, there is no charging infrastructure needed, while also allowing the driver to utilize familiar vehicle features. They provide a smooth transition toward cleaner driving by using existing practices and systems. This ease of transition improves the chances of many to adopt "greener" vehicles, which aid in achieving climate and air quality improvement goals.

The incorporation of internal combustion technology with electric propulsion makes HEVs easier to use. Their ability to reduce emissions and noise, while using existing refueling and maintenance networks, means they can have an immediate positive impact as the world transitions to fully electric vehicles in the future [8, 9].

1.6 Market Progression

Hybrid electric vehicles have come a long way over the past two decades. Advances in battery design, electronics, and control systems have made today's hybrids much more efficient. Thanks to these improvements, people now have a wide range of options, from compact city cars to larger SUVs and even commercial vehicles.

Hybrids usually cost a bit more than regular petrol or diesel cars, but many people find that the money they save on fuel balances things out in the long run. In some areas, government rebates or tax breaks can also help make that first purchase easier on your wallet. As more people become aware of environmental issues and as rules around car emissions get stricter, hybrids are quickly becoming a popular option.

Better hybrid technology, more public concern about the environment, and tougher emissions standards are all coming together to push even more drivers toward hybrids. These cars feel familiar to drive, but they help cut down on pollution, making them a smart and practical step toward cleaner transportation.

At the same time, the need to reduce car pollution is more urgent than ever. As cities get bigger and more people own cars, problems like dirty air and high fuel prices are only getting worse. The power grid is also feeling the strain, and energy costs can be unpredictable [10–12]. Fully electric cars are part of the answer, but they aren't always practical everywhere, especially in places where charging stations are hard to find or the power supply isn't reliable.

Fuel cells offer another promising solution. They make electricity from hydrogen and only give off water, so they don't add to air pollution. Because fuel cells don't rely on the main power grid, they can help set up charging stations even in areas where it's tough to build traditional infrastructure. With fuel cells in hybrid charging systems, it's possible to provide fast charging in cities and reliable charging in remote spots. This fits right in with the global push for cleaner, more flexible ways to travel and could play a big role in the future of transportation.

2 Hybrid Electric Vehicle Architectures

Hybrid electric vehicle (HEV) architectures have evolved to maximize efficiency, reduce emissions, and offer flexible performance across diverse driving scenarios. The three principal configurations—series, parallel, and series–parallel hybrids—each present unique design features, operational strategies, and trade-offs, and have been the subject of extensive research and real-world application.

2.1 *Series Hybrid Architecture*

In a series hybrid, the internal combustion engine (ICE) is decoupled from the wheels and operates primarily as a generator to produce electricity. This electricity either powers the electric motor, which is solely responsible for driving the wheels, or is stored in the battery for later use. The ICE can run at its most efficient operating point, independent of vehicle speed, which enhances fuel efficiency and reduces emissions. However, because the mechanical energy from the engine is converted to electrical energy and then back to mechanical energy at the wheels, there are additional conversion losses, making this architecture less efficient at high speeds or during sustained highway driving. Series hybrids are successful in urban environments with heavy traffic, where the electric motor's responsiveness and regenerative braking capabilities are maximized.

2.2 *Parallel Hybrid Architecture*

Parallel hybrid systems are the most common type found in today's hybrid electric vehicles. In this setup, both the gasoline engine and the electric motor are connected directly to the wheels. This means the car can be powered by the engine, the electric motor, or both at the same time, depending on what's needed. One big advantage of the parallel hybrid design is that it lets the engine send power straight to the wheels. This direct link is especially useful on highways, where it usually leads to better fuel economy compared to series hybrids, which rely only on the electric motor to drive the car. However, this arrangement does make the transmission system more complicated, since it has to smoothly balance power from both the engine and the motor in all sorts of driving situations.

There are several types of parallel hybrid setups, often labeled as P0, P1, P2, P3, and P4 [13]. These names refer to where the electric motor is placed in relation to the engine and the rest of the drivetrain. For example, in a P0 system, the motor is attached to the engine's crankshaft with a belt and mainly helps with starting, stopping, and regenerative braking. In P2 and P4 systems, the motor is positioned either between the engine and transmission or at the rear axle, which allows for stronger electric driving and better efficiency.

Each of these layouts comes with its own set of pros and cons—some are more efficient, some are simpler, and some cost less to make. This variety gives car makers the flexibility to design hybrids that fit different types of vehicles and meet the needs of a wide range of drivers.

2.3 *Series–Parallel (Power-Split) Hybrid Architecture*

The series–parallel, or power-split, hybrid system is designed to get the best out of both series and parallel hybrid setups. At the heart of this system is a planetary gear set, which lets the car switch smoothly between different modes depending on how it's being driven.

In city traffic or at low speeds, the hybrid typically runs in series mode. Here, the engine works mainly as a generator, providing electricity to the motor or charging the battery. This setup is great for stop-and-go driving, letting the electric motor do most of the work while the engine runs efficiently in the background.

What sets the series–parallel system apart is its flexibility. It automatically chooses the best power source based on your speed, battery level, and how much power you need at any moment. This means you get better fuel economy and a smoother ride, whether you're in city traffic or out on the open road.

A popular example of this approach is Toyota's Hybrid Synergy Drive, which uses a planetary gear system to balance power from the engine and electric motors. This technology not only helps save fuel but also delivers a responsive and comfortable driving experience.

2.4 *Mild, Full, and Plug-In Hybrids*

Hybrid electric vehicles differ in how much they use electric power, which shapes their efficiency and driving feel. Mild hybrids use small electric motors mainly to help the engine with tasks like start-stop and regenerative braking, but can't drive on electricity alone. Full hybrids have larger batteries and motors, letting them run briefly on electric power at low speeds and switch smoothly between engine and electric drive to save fuel, especially in city traffic.

Plug-in hybrids (PHEVs) go further, with bigger batteries that can be charged from an outlet. They can cover longer distances on electricity alone—often enough for daily commutes—before switching to hybrid mode when the battery runs low. This gives drivers the benefits of electric driving with the backup of a gasoline engine for longer trips.

By offering mild, full, and plug-in options, automakers can meet a range of needs and help more drivers transition to cleaner transportation.

2.5 Comparative Performance and Efficiency

Different hybrid electric vehicle (HEV) designs offer specific benefits depending on how and where they are used. Series hybrids are especially effective in city driving, as frequent stops and starts let the electric motor handle most of the work, while the engine mainly serves as a generator to recharge the battery. Parallel hybrids are more effective on highways, as both the engine and electric motor can work together to provide power directly to the wheels, improving fuel efficiency at higher speeds. Series–parallel (or power-split) hybrids combine elements of both systems, offering a balanced solution that adapts to both urban and highway driving, making them well-suited for mixed use. Recent literature highlights ongoing innovation in HEV architectures, such as the integration of ultracapacitors for improved energy storage, the use of 48 V mild hybrid systems for cost-effective efficiency gains, and the incorporation of renewable energy sources like solar panels in hybrid solar vehicles. These advancements continue to expand the potential applications and benefits of hybrid technology in the pursuit of cleaner, more sustainable transportation.

3 Fuel Cell Technology in Hybrid Systems

In a fuel cell, the fundamental operation is the conversion of hydrogen's chemical energy directly into electricity. Fuel cells electrochemically convert the chemical energy stored in hydrogen (**H_2**) directly into electricity, with water and heat as the only byproducts. Unlike combustion-based engines, this process avoids inefficient thermal cycles, offering higher energy efficiency.

The overall reaction consists of two half-reactions occurring at the anode and cathode:

Anode (Hydrogen Oxidation):

$$H_2 \rightarrow 2H^+ + 2e^-$$

Hydrogen molecules split into protons (H^+) and electrons (e^-). Protons pass through the membrane, while electrons flow through an external circuit, generating an electric current.

Cathode (Oxygen Reduction):

$$12O_2 + 2H^+ + 2e^- \rightarrow H_2O_21O_2 + 2H + 2e^- \rightarrow H_2O$$

Oxygen from the air combines with protons and electrons to form water.

Net Reaction:

$$H_2 + 12O_2 \rightarrow H_2O + \text{Electrical Energy} + \text{Heat}$$

$$H_2 + 21O_2 \rightarrow H_2O + \text{Electrical Energy} + \text{Heat}$$

In hybrid systems, the above process consumes minimum energy as there is no combustion. Also, if we stack multiple cells, the voltage and power for vehicular use are increased. Also, a fuel cell provides a steady base power, which can optimize the efficiency of the vehicle.

3.1 Main Component of a Fuel Cell

A fuel cell consists of several critical components that work together to convert hydrogen and oxygen into electricity, with water and heat as the only byproducts. As illustrated in Fig. 2, the anode (fuel inlet) facilitates hydrogen oxidation $H_2 \rightarrow 2H^+ + 2e^-$, where protons ($H^+$) pass through the polymer electrolyte membrane (PEM) while electrons (e^-) flow through an external circuit, generating an electric current. The cathode (air inlet) combines oxygen ($2O_2$), protons, and electrons to produce water ($2H_2O$) as shown in Fig. 2. The PEM selectively allows proton conduction but blocks electron transfer, ensuring efficient energy conversion. Excess fuel and byproducts (water/heat) are safely expelled. Key supporting elements include gas diffusion layers (for even reactant distribution) and bipolar plates (for structural integrity and current collection). Together, these components enable clean, high-efficiency power generation.

3.2 Advantages of Fuel Cells

Fuel cells offer transformative benefits for hybrid electric vehicle (HEV) charging infrastructure by addressing critical challenges in sustainability, efficiency, and grid independence. They operate by electrochemically converting hydrogen into electricity, producing only water vapor and heat as byproducts. This process results in zero tailpipe emissions, eliminating pollutants such as carbon dioxide, nitrogen oxides, and particulate matter, which significantly contribute to air quality improvement in urban environments and support global decarbonization efforts. In terms of efficiency, fuel cells outperform conventional combustion engines, achieving 50–60% energy conversion rates, and when combined with regenerative braking in HEVs, can push total system efficiency beyond 80%, thereby reducing overall energy waste. Furthermore, fuel cells seamlessly integrate with renewable hydrogen sources, where hydrogen produced through electrolysis powered by solar or wind energy creates a

Fig. 2 Fuel cell [8]

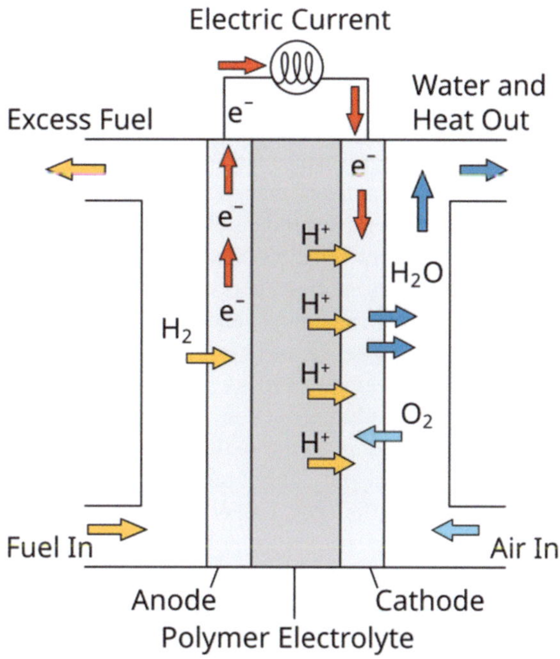

clean, closed-loop energy cycle. This not only leverages excess renewable energy for storage but also mitigates intermittency issues commonly associated with solar and wind power. Fuel cell systems also provide substantial benefits in terms of grid independence and scalability [14], allowing charging stations to operate off-grid, which is particularly advantageous for remote or disaster-stricken areas. Fuel cells provide steady power to help balance the electrical grid, especially during peak demand, and work well alongside battery storage. Fuel cell vehicles also refuel quickly—usually in just a few minutes—similar to gasoline cars, which solves the long charging times of battery electric vehicles. In hybrid setups, combining fuel cells and batteries boosts both driving range and flexibility. While challenges like hydrogen production costs and infrastructure remain, new technologies such as solar-powered hydrogen and supportive government policies are making progress [19–24]. As these improvements continue, fuel cells are becoming an essential part of future hybrid vehicle charging systems, combining environmental sustainability with the performance and reliability required for the widespread adoption of clean mobility solutions. Apart from this, other areas, including power quality and grid-connected EV integrating EV motors, can be explored [15–19].

4 Fuel Cell Hybrid Electric Vehicles (FCHEVs)

A Fuel Cell Hybrid Electric Vehicle (FCHEV) [25–27] power train strategically combines a fuel cell, a battery, and optionally an ultracapacitor to optimize energy usage, enhance performance, and extend system longevity as shown in Table 1. The fuel cell functions as the primary energy source, converting hydrogen and oxygen into electricity to power the vehicle's electric motor and recharge the battery. The battery in a fuel cell hybrid vehicle acts as a backup energy source, providing extra power during acceleration, storing energy from regenerative braking, and helping out when the fuel cell can't react quickly enough. Some designs also include an ultracapacitor, which can quickly deliver or absorb bursts of energy—ideal for sudden acceleration or braking. A power electronics controller manages how energy moves between the battery, fuel cell, ultra capacitor, and electric motor, making sure everything runs efficiently. Depending on the vehicle's needs, the power train can be set up in series, parallel, or power-split configurations. This setup lets each part work at its best, boosting overall efficiency, fuel economy, and driving performance. By sharing the energy load, the system also helps components last longer and increases the vehicle's range. Overall, this hybrid approach makes fuel cell vehicles more flexible, efficient, and reliable for sustainable transportation.

4.1 Energy Management Strategies for Fuel Cell Hybrid Electric Vehicles (FCHEVs)

Effective energy management is essential for fuel cell hybrid electric vehicles (FCHEVs). These systems control how power is shared between the fuel cell and storage units like batteries, aiming to use less hydrogen, keep the battery in a safe charge range, and make components last longer—all while meeting the driver's needs.

One advanced method, Model Predictive Control (MPC), uses route and traffic data to predict power needs. This helps keep the battery charge steady (usually between 50 and 70%) and the fuel cell running efficiently. Studies show MPC can cut hydrogen use by up to 15% and reduce stress on the fuel cell.

Table 1: Power Sources ...

Power source	Role in FCHEV powertrain	Key benefit
Fuel cell	Main energy provider (hydrogen to electricity)	Long-range, low emissions
Battery	Stores/regenerates energy, supports acceleration	Power smoothing, regenerative braking
Ultracapacitor	Rapid charge/discharge for peak power events	High power density, buffering

Table 2: Performance Strategies ….

Strategy	Hydrogen savings (%)	FC degradation reduction	Computational complexity
MPC	5–15	Moderate	High
PMP-DCR	~ 14	High (93% versus rule-based)	Moderate
Fuzzy logic	5–10	Low	Low
ECMS	10–12	Moderate	Moderate

Another technique, Pontryagin's Minimum Principle (PMP), adapts to driving conditions—city, suburban, or highway—to fine-tune energy use. Research finds PMP can save nearly 14% on hydrogen and cut fuel cell start-stop cycles by 30%, which is especially useful for buses and other vehicles with regular routes. Simple controllers like fuzzy logic and rule-based systems are still common in commercial vehicles because they're reliable and easy to use. They don't need much computing power and can handle unpredictable situations.

The Equivalent Consumption Minimization Strategy (ECMS) looks at battery use in terms of hydrogen savings, allowing for real-time energy adjustments. ECMS works even better when paired with ultracapacitors, which help handle sudden power demands.

Recent strategies also focus on protecting key parts. They avoid quick changes in fuel cell output and keep the battery from being overworked or overheated, which helps everything last longer.

Overall, these smart management strategies can boost system efficiency by up to 25%, reduce hydrogen use by as much as 30%, and help batteries and fuel cells last longer. This makes them crucial for building reliable and eco-friendly fuel cell vehicles. A comparison is drawn based on the performance of highlighted strategies in Table 2.

4.2 Integrating Fuel Cells into EV Charging Infrastructure

Integrating hydrogen fuel cells into electric vehicle (EV) charging networks offers a promising way to address grid constraints and boost energy resilience. By using hydrogen produced from renewable sources, these systems can operate independently of the main power grid, making them ideal for high-traffic areas like airports, commercial centers, and remote locations. Stationary fuel cells provide a steady and reliable power supply for EV chargers, which are especially valuable in rural regions, national parks, and along highways where expanding the electrical grid is expensive or unfeasible. This approach not only supports sustainable transportation but also ensures that EV charging remains accessible and dependable, even in off-grid or hard-to-reach areas. For instance, an 18,000-gallon liquid hydrogen storage system can power over

1,000 EV charges, making it a scalable solution for underserved regions. Additionally, fuel cells supplement grid power during peak demand, preventing blackouts and stabilizing energy costs by avoiding volatile electricity pricing. The environmental benefits of this approach hinge on the adoption of green hydrogen, produced via electrolysis powered by excess solar or wind energy, which ensures a zero-emission charging infrastructure. However, while 95% of current hydrogen production relies on fossil fuels, scaling up renewable hydrogen infrastructure remains essential to fully realize its decarbonization potential.

Commercial applications and pilot projects demonstrate the viability of hydrogen-powered EV charging, particularly in high-traffic and mission-critical environments. Airports, such as Los Angeles International (LAX), have deployed hydrogen fuel cell stations to support electric ground vehicles—baggage tugs, shuttles, and service fleets—without disrupting operations. Similarly, logistics giants like Amazon and FedEx utilize hydrogen-powered depots for fast, off-grid charging of delivery vans, circumventing expensive grid upgrades in urban centers. Partnerships between technology firms, such as ABB and AFC Energy, have led to the development of high-power, off-grid EV chargers using industrial-grade hydrogen, reducing costs compared to ultra-pure fuel cell alternatives. Despite these advancements, challenges persist, including high upfront costs for electrolyzers and hydrogen storage infrastructure. Nevertheless, the scalability, resilience, and dual compatibility of hydrogen fuel cells—supporting both battery EVs and fuel cell vehicles—position them as a cornerstone of future-proof, sustainable charging networks. As renewable hydrogen production scales, fuel cells will play an increasingly pivotal role in building a resilient, zero-emission transportation ecosystem.

4.3 Benefits and Challenges of Integration

The transition to fuel cell electric vehicles (FCEVs) and hydrogen-powered charging infrastructure presents both significant opportunities and formidable challenges that will shape the future of clean transportation. While FCEVs currently demonstrate clear advantages in heavy-duty and long-range applications—particularly in commercial vehicles and industrial equipment—their widespread adoption requires coordinated efforts across technological, economic, and policy domains. The environmental benefits are undeniable: FCEVs produce zero tailpipe emissions, with only water vapor as a byproduct, and when powered by green hydrogen, they offer a truly sustainable alternative to fossil fuels. Their operational advantages, including rapid refueling (~5 min) and extended range (>300 miles), address critical limitations of battery electric vehicles (BEVs), particularly for fleet operators and long-haul transportation. However, the current hydrogen infrastructure gap remains a major barrier, with limited production capacity, storage solutions, and distribution networks hindering scalability.

To get the most out of fuel cell electric vehicles (FCEVs), a few important steps need to be taken. First, we need to make more hydrogen using clean energy since most hydrogen today still comes from fossil fuels. Newer electrolyzers and more renewable power are helping to lower the cost and make hydrogen production greener. It's also important for governments and businesses to help out by funding hydrogen stations and offering tax breaks to people who buy fuel cell cars.

As these changes happen, FCEVs will likely be used alongside battery electric vehicles (BEVs), with each type fitting different driving needs and places. If progress continues and policies stay supportive, FCEVs could cost about the same as regular cars by 2030. This would make them a bigger part of the move to cleaner transportation and help cut down on pollution worldwide.

5 Conclusion

This chapter studies how using fuel cells can make charging systems for hybrid electric vehicles (HEVs) stronger and more reliable. By mixing fuel cells with batteries and ultracapacitors, these systems can work even when the power grid is weak or not available, deal with changes in solar or wind power, and supply enough energy for busy places like airports or shopping centers. Based on research, real-world examples, and what people in the industry have learned, the chapter showed that these hybrid systems can make charging more dependable, cut down on pollution, and give good solutions for charging electric cars—especially in areas where it's hard or expensive to build new power lines. The main findings highlight that fuel cells make it possible to set up robust, off-grid charging stations, which are especially valuable for remote areas, commercial fleets, and sites with heavy traffic. Using hydrogen produced from renewable sources, these systems can deliver truly zero-emission charging. However, challenges remain, including the cost of hydrogen, the need for better infrastructure, and improving the lifespan of fuel cells.

Comparisons show that while battery electric vehicles (BEVs) are well-suited for city driving and lighter vehicles, fuel cell hybrids have clear advantages for long-distance travel and heavy-duty use. Rather than replacing BEVs, fuel cell vehicles are expected to complement them, each serving different needs within the growing clean transport sector.

The synthesis of multidisciplinary research in this chapter serves as a foundational resource for new research. Ultimately, this chapter positions fuel cell-integrated charging systems as a pivotal enabler of next-generation transportation ecosystems, bridging the gap between renewable energy adoption and practical, large-scale EV deployment. By translating fragmented research into actionable insights, it aims to accelerate progress toward a decarbonized, resilient, and inclusive energy future.

References

1. Chen J, He H, Wang Y-X, Quan S, Zhang Z, Wei Z, Han R (2024) Research on energy management strategy for fuel cell hybrid electric vehicles based on improved dynamic programming and air supply optimization. Energy 300(C)
2. Tie SF, Tan CW (2013) A review of energy sources and energy management system in electric vehicles. Renew Sustain Energy Rev 20:82–102. ISSN 1364-0321. https://doi.org/10.1016/j.rser.2012.11.077
3. Fan L, Tu Z, Chan SH (2021) Recent development of hydrogen and fuel cell technologies: a review. Energy Rep 7:8421–8446. ISSN 2352-4847. https://doi.org/10.1016/j.egyr.2021.08.003
4. Anchasa Pramuanjaroenkij S (2023) The fuel cell electric vehicles: the highlight review. Int J Hydrog Energy 48(25):9401–9425. ISSN 0360-3199. https://doi.org/10.1016/j.ijhydene.2022.11.103
5. Zhao Z, Lee CKM, Ren J (2024) A two-level charging scheduling method for public electric vehicle charging stations considering heterogeneous demand and nonlinear charging profile. Appl Energy 355:122278
6. Bukya M, Meenakshi Reddy R, Doddipatla A, Kumar R, Mathur A, Gupta M, Garimella A (2024) Electro-thermal performance evaluation of a prismatic battery pack for an electric vehicle. High-Temp Mater Process 43(1):20220311. https://doi.org/10.1515/htmp-2022-0311
7. Zhuang W, Eben SL, Zhang X, Kum D, Song Z, Yin G, Ju F (2020) A survey of powertrain configuration studies on hybrid electric vehicles. Appl Energy 262:114553. ISSN 0306-2619. https://doi.org/10.1016/j.apenergy.2020.114553
8. Juan M. Ramirez et al, 12 - Study of harmonics in linear, nonlinear nonsinusoidal electrical circuits by geometric algebra,Editor(s): Emilio Barocio Espejo, Felix Rafael Segundo Sevilla, Petr Korba, Monitoring and Control of Electrical Power Systems Using Machine Learning Techniques, Elsevier, 2023,Pages 289-308,ISBN 9780323999045, https://doi.org/10.1016/B978-0-32-399904-5.00018-1.
9. Sharma S, Kundu S, Singh S, Singh J, Singh S (2025) Harmonics mitigation and control of grid-connected PV system under intermittent conditions. In: Kumar A, Singh SN, Kumar P (eds) Energy management in renewable sources integrated system. ICGEST 2023. Lecture Notes in Networks and Systems, vol 1236. Springer, Singapore. https://doi.org/10.1007/978-981-96-1012-9_26
10. S. Kundu, N. Sundriyal, A. Dixit, S. Singh, S. Kadiyan and R. Kundu, "Improving Power Quality in Grid-Connected Photovoltaic Systems with FLL-Based Control for Nonlinear Demands," 2025 IEEE International Conference on Computer, Electronics, Electrical Engineering & their Applications (IC2E3), Srinagar Garhwal, India, 2025, pp. 1-6, doi: 10.1109/IC2E365635.2025.11166983.
11. Boddapati V, Kumar R, Arul Daniel A, Padmanaban S (2022) Design and prospective assessment of a hybrid energy-based electric vehicle charging station. Sustain Energy Technol Assess 53:102389
12. Kundu S, Giri AK, Singh M, Singh S (2025) Effective multiport battery charging facility, operation and control in hybrid isolated microgrid system. In: Giri AK, Singh M (eds) Electric vehicle charging infrastructures and its challenges. Studies in Infrastructure and Control. Springer, Singapore. https://doi.org/10.1007/978-981-96-0361-9_11
13. Sundriyal N, Ramirez JM, Corrachano EB (2023) A comparison of geometric algebra and harmonic domain for linear circuit analysis. In: AGACSE-2021, mathematics for application, vol 1. https://doi.org/10.13164/ma.2023.08
14. Sundriyal N, Ramirez JM, Saini P, Dixit A (2023) Analytical voltage sag and swell characterization using differential geometry. In: 2023 International conference on computer, electronics & electrical engineering & their applications (IC2E3), Srinagar Garhwal, India, pp 1–5. https://doi.org/10.1109/IC2E357697.2023.10262816

15. Sundriyal N, Sharma V, Kumar A, Dixit A (2018) Cogent fault diagnosis of photovoltaic-fed single-phase supply system. In: 2018 IEEE 8th power india international conference (PIICON), Kurukshetra, India, pp 1–5. https://doi.org/10.1109/POWERI.2018.8704386
16. Gutiérrez-Torres DA, Ramìrez JM, García JML, Sundriyal N, Bhatt A (2024) Implementation of a direct AC-AC converter based wind energy conversion system. In: 2024 International conference on computer, electronics, electrical engineering & their applications (IC2E3), Srinagar Garhwal, Uttarakhand, India, pp 1–5. https://doi.org/10.1109/IC2E362166.2024.10827016
17. Kundu S, Sundriyal N, Singh N, Kadiyan S, Singh S, Ramirez JM (2025) Control strategy for a single-phase grid-connected PV system with bidirectional EV charging support. In: Fotis G, Gupta S, Varshney T, Dhar S, Gulzar M (eds) Modern computing technologies for EV efficiency and sustainable energy integration. IGI Global Scientific Publishing, pp 179–198. https://doi.org/10.4018/979-8-3373-2382-4.ch007
18. Kundu S, Singh M, Giri AK (2024) Synchronization and control of WECS-SPV-BSS-based distributed generation system using ICCF-PLL control approach. Electr Power Syst Res 226:109919. ISSN 0378-7796. https://doi.org/10.1016/j.epsr.2023.109919
19. Kundu S, Singh M, Giri AK (2024) SPV-wind-BES-based islanded electrical supply system for remote applications with power quality enhancement. Electr Eng 106:279–294. https://doi.org/10.1007/s00202-023-01979-0
20. Kundu S, Singh M, Giri AK et al (2024) Robust and fast control approach for islanded microgrid system and EV charging station applications. Electr Eng. https://doi.org/10.1007/s00202-024-02291-1
21. Kundu S, Giri AK, Singh M et al (2024) Adaptive control strategy for isolated renewable energy-based generation system with intermittency. Electr Eng. https://doi.org/10.1007/s00202-024-02454-0
22. Kundu S, Giri AK, Kadiyan S, Singh S (2025) Power management and adaptive control method for a standalone distributed generation system under fault, DC-offset, and dynamic environments. Int J Ambient Energy 46(1). https://doi.org/10.1080/01430750.2025.2452245
23. Vendoti S, Muralidhar M, Kiranmayi R (2021) Techno-economic analysis of off-grid solar/wind/biogas/biomass/fuel cell/battery system for electrification in a cluster of villages by HOMER software. Environ Dev Sustain 23:351–372
24. Kundu S, Giri AK, Kadiyan S (2025) An adaptive mixed-step size normalized least means fourth control approach for stand-alone power generation system considering dynamic conditions. IEEE J Emerg Sel Top Power Electron 13(2):2422–2429. https://doi.org/10.1109/JESTPE.2024.3516878
25. Verma SK, Kundu S, Singh S (2022) Design and techno-economic analysis of electric vehicle charging station for a railway metro station parking area. In: Mahajan V, Chowdhury A, Padhy NP, Lezama F (eds) Sustainable technology and advanced computing in electrical engineering. Lecture Notes in Electrical Engineering, vol 939. Springer, Singapore. https://doi.org/10.1007/978-981-19-4364-5_67
26. Improved dynamic performance of SEIG-based wind energy systems for off-grid applications,Electric Power Systems Research,Volume 251,2026,112198,ISSN 0378-7796. https://doi.org/10.1016/j.epsr.2025.112198
27. Gupta, Sandeep, et al. Sustainable automation: Balancing efficiency and environmental goals. Optimizing automation in engineering with energy systems and communication networks, edited by Vipin Balyan, et al., IGI Global Scientific Publishing, 2026, pp. 93-120. https://doi.org/10.4018/979-8-3373-2737-2.ch004

Chapter 10
Renewable Energy-Based EV Battery Swapping Stations: Opportunities and Future Directions

Prasanta Kumar Mohanty, Rudranarayan Pradhan, Premalata Jena, and Narayana Prasad Padhy

Abstract The global shift toward sustainable transportation is accelerating with the widespread adoption of EVs, which have surpassed 26 million units globally as of 2023. Despite this progress, challenges such as prolonged charging times, limited charging infrastructure, and heavy grid dependence continue to hinder large-scale EV integration. Battery swapping has emerged as a viable alternative, offering rapid energy replenishment while decoupling charging from vehicle downtime. Unlike traditional charging, battery swapping can reduce peak grid load impact by up to 50% compared to fast charging systems, significantly alleviating stress on power networks. This chapter investigates the integration of renewable energy sources—including solar, wind, and hybrid systems—into EV battery swapping stations to improve environmental sustainability, enhance grid independence, and increase operational efficiency. A detailed examination of system architecture, energy storage management, power electronics interfaces, and smart energy management systems is presented. Additionally, the chapter reviews global initiatives, real-world case studies, and techno-economic assessments supporting the scalability of renewable-powered battery swapping infrastructure. Key performance indicators such as impact ratings of 9/10 for reduced downtime and cost savings, and 8/10 for battery life extension and grid load management further validate the operational benefits of this model. Key barriers such as renewable intermittency, lack of standardization, battery ownership complexities, and regulatory constraints are critically analyzed. The chapter

P. K. Mohanty · P. Jena · N. P. Padhy
Department of Electrical Engineering, Indian Institute of Technology Roorkee, Roorkee, India
e-mail: pkumarmohanty@ee.iitr.ac.in

P. Jena
e-mail: premalata.jena@ee.iitr.ac.in

N. P. Padhy
e-mail: nppadhy@ee.iitr.ac.in

R. Pradhan (✉)
School of Electrical Sciences, Odisha University of Technology and Research, Bhubaneswar, Odisha, India
e-mail: rudranarayan@outr.ac.in

S. Kundu et al. (eds.), *Hybrid Renewable Power Infrastructure for Sustainable Electric Vehicle Development*, Studies in Infrastructure and Control,
https://doi.org/10.1007/978-981-95-1323-9_10

concludes with strategic research directions and emerging innovations necessary to realize a green, intelligent, and scalable EV battery swapping ecosystem.

Keywords Battery swapping station (BSS) · Electric vehicles (EVs) · Renewable energy integration · Smart energy management · Vehicle-to-grid (V2G) · System architecture · Power electronics · Sustainable mobility

1 Introduction

1.1 Motivation for Sustainable Transportation

The global transportation sector accounts for nearly 24% of direct CO_2 emissions from fuel combustion, with road transport alone contributing approximately 15% of total global emissions [1]. As of 2023, the number of vehicles in operation worldwide surpassed 1.45 billion, with projections suggesting it could approach **2 billion by 2050** if current growth trends persist. This increase, largely driven by urbanization and economic development in emerging markets, has escalated both energy consumption and environmental degradation—making transportation a key contributor to climate change and public health risks.

Internal combustion engine (ICE) vehicles, which continue to dominate the global fleet, emit an estimated **2.3 kg of CO_2** per liter of gasoline combusted. In addition to greenhouse gases, ICE vehicles also release harmful air pollutants such as nitrogen oxides (NO_x) and particulate matter ($PM_{2.5}$), which are strongly associated with respiratory and cardiovascular diseases [2]. According to the World Health Organization, outdoor air pollution causes over **4.2 million premature deaths annually**, with transportation being a leading contributor in urban centers.

In response, electrification of transport has emerged as a viable pathway toward sustainable mobility. Electric vehicles (EVs), when powered by clean electricity, offer considerable environmental advantages—emitting less than half the CO_2 per kilometer compared to ICE counterparts. Their inherent energy efficiency is another major advantage, with EVs converting over **77%** of the electrical energy from the grid into motion, far exceeding the **12–30%** efficiency range of ICEs. Figure 1 visualizes several key sustainability metrics within the EV ecosystem. Figure 1(a) illustrates the exponential increase in global EV stock over the past decade, while Fig. 1(b) highlights the superior energy efficiency of EVs over ICE vehicles. Figure 1(c) shows the potential for significant lifecycle CO_2 emission reductions when EVs are powered by renewable sources. Lastly, Fig. 1(d) confirms that road transport remains the largest contributor to CO_2 emissions within the broader transport sector.

The integration of renewable energy—particularly solar and wind—into EV charging and battery swapping infrastructure further enhances environmental performance by reducing dependence on fossil-fuel-derived electricity. This alignment with global decarbonization frameworks, such as the Paris Agreement, strengthens

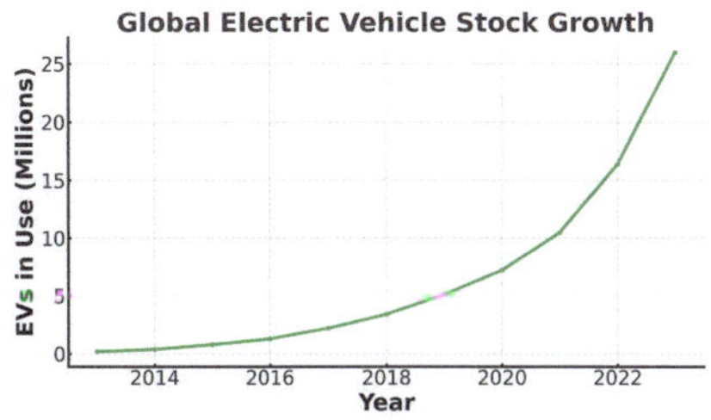

(a) EV Stock Growth (2013–2023)

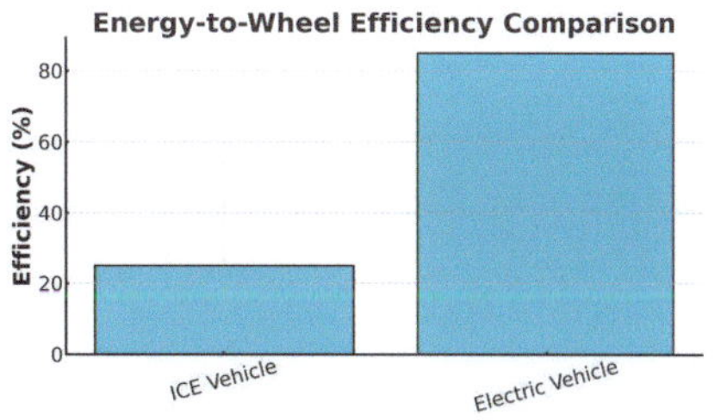

(b) Energy Efficiency: ICE vs EV

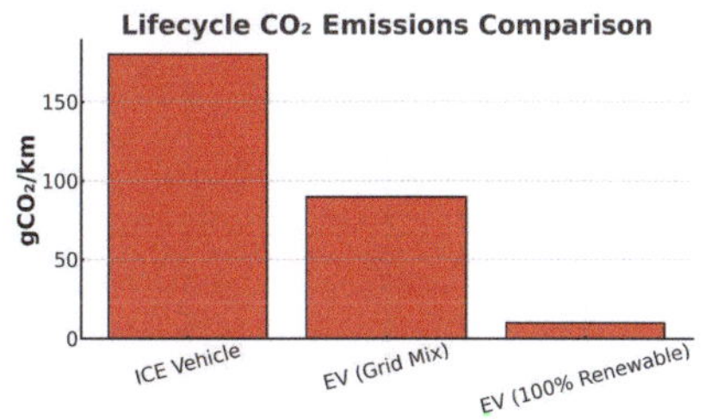

(c) Lifecycle CO_2 Emissions

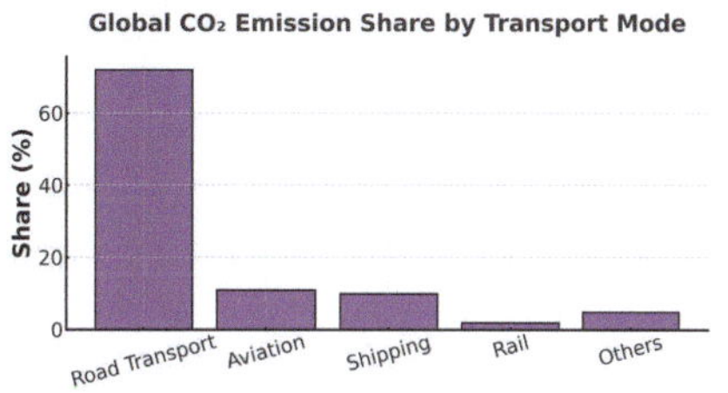

(d) Transport Sector CO_2 Emission Share

Fig. 1 Key sustainability metrics for EVs: (**a**) global EV stock trends, (**b**) energy conversion efficiency comparison, (**c**) lifecycle CO_2 emissions, and (**d**) transport sector emission breakdown

national efforts to meet emission reduction targets, improve air quality, and reduce reliance on oil imports. From an economic perspective, the EV market is projected to exceed **$950 billion by 2030**, fueled by technological innovation, declining battery costs, and favorable regulatory environments. In parallel, the renewable energy sector continues to expand rapidly, with solar and wind accounting for over **70% of new global electricity capacity additions in 2022**. These trends collectively reinforce the urgency and feasibility of transitioning to electrified, renewable-powered mobility. In this context, battery swapping stations (BSS) offer a scalable and decentralized energy delivery solution—enabling fast, reliable, and sustainable EV refueling, particularly in high-utilization applications such as urban fleets and public transport.

1.2 Limitations of Conventional EV Charging

While EVs are widely recognized as a key solution for decarbonizing the transportation sector and reducing urban air pollution, their widespread adoption continues to be constrained by several inherent limitations of conventional EV charging infrastructure. Chief among these is the issue of long charging times—standard Levels 1 and 2 chargers can require several hours to fully recharge an EV battery. Even fast DC chargers, while quicker, still fall short of matching the convenience and speed of internal combustion engine (ICE) vehicle refueling, which typically takes only a few minutes [3–5]. This time inefficiency is a significant deterrent for potential users and undermines the practicality of EVs in time-sensitive use cases.

A further impediment is the limited and uneven distribution of public charging infrastructure. In many developing countries and rural areas, charging points are sparse or non-existent, contributing to the phenomenon of "range anxiety"—the fear of depleting battery charge before reaching a charging station [6–8]. This infrastructural gap not only discourages EV adoption but also exacerbates inequities in access to clean mobility options. Moreover, as the number of EVs increases, so does the aggregate demand placed on the power grid. Uncontrolled and simultaneous charging—especially during peak hours—can cause significant stress on local distribution networks, potentially leading to grid instability and increased operational costs for utilities [9].

Compounding these issues is the unidirectional nature of most conventional charging systems, which limits the integration of vehicle-to-grid (V2G) technologies. V2G capabilities, where EVs can discharge energy back to the grid, are crucial for providing ancillary services such as frequency regulation, peak shaving, and emergency backup. The absence of such functionalities restricts the potential of EVs to function as distributed energy resources (DERs), thereby missing an opportunity to enhance grid flexibility and resilience [10]. Additionally, in many regions, conventional EV charging stations rely heavily on electricity derived from fossil fuels. This undermines the environmental benefits of EVs, particularly when carbon-intensive electricity is used for charging. The lack of synchronization with renewable energy generation further dilutes the sustainability case for EVs [11–13].

As highlighted in [1], the integration of renewable energy sources—such as solar PV—into conventional charging models is often poorly executed, owing to the absence of advanced energy management and storage strategies. This not only increases operational costs but also results in underutilization of clean energy. A detailed summary of these limitations and their associated impact domains is provided in Table 1. The table emphasizes how these shortcomings span across critical dimensions, including user experience, grid stability, and environmental footprint.

To address these challenges, battery swapping stations (BSSs) have emerged as a compelling alternative. By decoupling battery charging from vehicle usage, BSSs drastically reduce wait times, enable off-peak or renewable-integrated charging, and allow for more effective energy distribution planning. As demonstrated in recent studies [14–18], the integration of BSSs with renewable energy sources and smart grid functionalities presents a pathway to overcoming the inefficiencies of traditional charging methods. These systems not only enhance service quality but also contribute to grid load balancing and carbon emission reductions, making them a critical component of next-generation EV infrastructure.

Table 1 Key limitations of conventional EV charging infrastructure and their impact domains

Limitation	Description	Impact domain
Long charging time	Standard Level 1/2 chargers take hours, even fast DC chargers are slower than ICE refueling	User experience, charging convenience
Limited charger availability	Sparse and unevenly distributed public charging infrastructure, especially in rural or developing regions	Accessibility, range anxiety, EV adoption
Grid load stress	High EV adoption without load balancing strategies causes peak demand and grid congestion	Grid stability, operational costs
Lack of V2G integration	Most charging is unidirectional, missing the opportunity for grid support via energy feedback	Ancillary services, grid flexibility
Fossil-based grid dependency	Charging often relies on non-renewable sources, limiting environmental benefits of EVs	Environmental footprint, carbon emissions
Battery degradation from fast charging	High-current fast charging accelerates battery wear	Battery lifespan, maintenance costs

1.3 Emergence of Battery Swapping Technology

Battery swapping has emerged as an innovative and practical alternative to conventional plug-in EV charging, addressing several of the critical limitations associated with traditional charging infrastructure. In a typical battery swapping model, a depleted battery is quickly replaced with a fully charged one at a BSS, significantly reducing vehicle downtime. The entire process can be completed within minutes—comparable to the refueling time of internal combustion engine vehicles—thus eliminating one of the primary barriers to EV adoption: long charging durations [3, 6]. This rapid turnaround is especially advantageous for high-utilization vehicles such as taxis, ride-sharing fleets, public buses, and delivery vans, where operational time is directly tied to service profitability. Battery swapping also decouples charging from driving, enabling operators to recharge batteries during off-peak hours and redistribute them when demand arises. This operational flexibility not only enhances user convenience but also contributes to grid stability by smoothing load profiles and avoiding stress during peak periods [9, 10]. The integration of intelligent scheduling and demand-side optimization in BSS infrastructure facilitates advanced load forecasting and effective energy dispatch, making the system more responsive to both user needs and grid requirements.

One of the key transformative aspects of battery swapping lies in the shift from individual to centralized battery ownership and management. In most swapping business models, batteries are owned and maintained by service providers rather than end-users. This model simplifies concerns related to battery degradation, warranty, and replacement, while also enabling real-time health monitoring and predictive maintenance through centralized diagnostics [11]. Such aggregation allows for more efficient lifecycle management, fosters the development of standardized battery formats, and facilitates recycling and second-life usage-critical components for building a sustainable and circular battery economy [14]. Moreover, BSSs are inherently more compatible with renewable energy integration compared to conventional plug-in stations. By aligning charging schedules with periods of high renewable energy availability—such as peak solar generation—BSSs can absorb excess electricity that would otherwise be curtailed. Several studies have demonstrated the potential of solar-powered or hybrid renewable BSS setups that combine photovoltaic systems with energy storage and grid interaction mechanisms [1, 15]. These stations can act as distributed energy resources (DERs), offering ancillary services such as frequency regulation, peak shaving, and voltage support to the grid.

The strategic deployment of battery swapping systems also enhances infrastructure modularity and mobility. Innovative concepts such as Mobile Swapping Stations (MSSs) and modular nanogrid-based BSS clusters have been shown to improve service coverage and flexibility, particularly in areas with fluctuating demand or weak grid connections [19]. With growing global momentum, particularly in densely populated urban environments and logistics-intensive sectors, battery swapping is now recognized as a scalable, grid-friendly, and renewable-compatible technology that holds the potential to significantly accelerate the global transition to electric mobility.

1.4 Role of Renewable Energy Integration

Integrating renewable energy sources into EV infrastructure plays a pivotal role in achieving a truly sustainable transportation ecosystem. While EVs offer zero tailpipe emissions and improved energy efficiency, their environmental impact heavily depends on the source of electricity used for charging. In regions where the electricity grid is predominantly powered by fossil fuels, the net reduction in greenhouse gas (GHG) emissions from EVs can be marginal. Hence, coupling EV systems with clean, renewable energy sources—such as solar photovoltaics (PV), wind turbines, and hybrid renewable configurations—can drastically enhance their environmental performance [15, 20]. BSSs integrated with renewable energy present a powerful synergy. These stations can leverage solar or wind power to charge a fleet of batteries during daylight or high-wind conditions, storing the energy in high-capacity battery banks for later use [14, 21]. This not only reduces dependence on the conventional grid but also improves energy resilience, particularly in areas with unreliable or constrained grid infrastructure [11]. Moreover, renewable energy-powered BSS support advanced energy management strategies such as load shifting, peak shaving, and participation in demand response programs. When combined with smart grid technologies and battery-to-grid (B2G) capabilities, these systems can act as decentralized energy hubs that both consume and supply electricity, thereby enhancing grid stability and supporting the integration of intermittent renewable sources [9, 10, 22]. From an economic standpoint, utilizing renewable energy reduces long-term operational costs by mitigating fuel price volatility and lowering grid electricity expenses. Studies show that PV-powered and hybrid systems can lead to significant lifecycle cost savings and reduced energy procurement from fossil-based grids [23]. These benefits align with global decarbonization goals and national energy transition strategies. Thus, the integration of renewable energy into BSS infrastructure is not merely an add-on but a transformative step toward achieving clean, cost-effective, and energy-secure electric mobility.

2 Overview of Battery Swapping Technology

2.1 Battery Swapping Versus Conventional Charging

EVs can be recharged using two primary methods: conventional plug-in charging and battery swapping. Each method offers distinct operational advantages and constraints.

2.1.1 Conventional Charging

Conventional charging involves connecting an EV to a power source via a charging cable. Based on voltage and power levels, conventional charging is typically categorized into:

- **Level 1 Charging**: Uses standard 120 V household outlets, requiring 8–12 h for a full charge.
- **Level 2 Charging**: Utilizes 240 V outlets, reducing charge time to 4–6 h.
- **DC Fast Charging**: Employs high-power direct current, achieving 80% charge in approximately 30 minutes.

Despite infrastructure growth, conventional charging presents challenges such as long wait times, grid stress during peak demand, and large spatial and network planning requirements.

2.1.2 Battery Swapping

Battery swapping replaces a depleted battery with a fully charged one at a BSS. The process typically takes less than five minutes, enabling time efficiency comparable

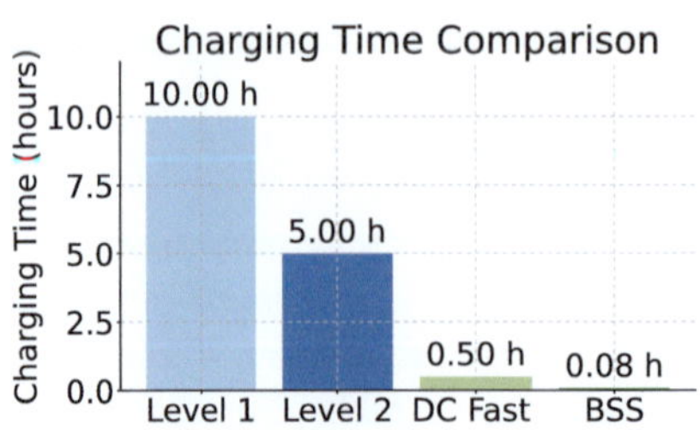

(a) Charging Time Comparison

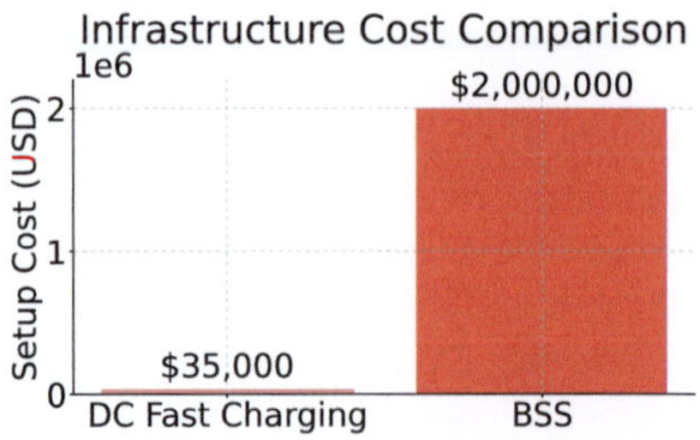

(b) Infrastructure Cost Comparison

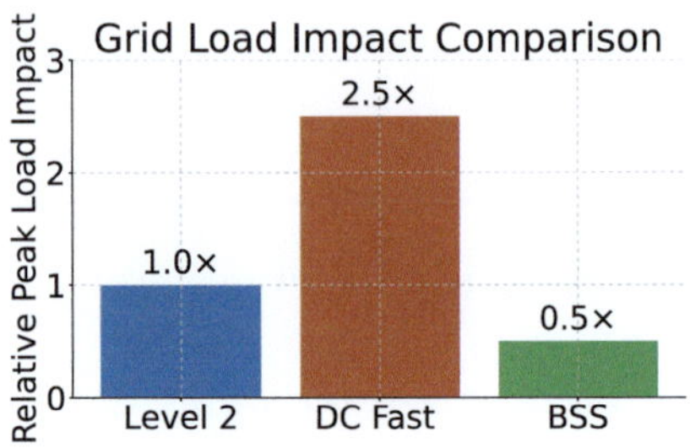

(c) Grid Load Impact Comparison

Fig. 2 Quantitative comparison of EV energy replenishment methods: **a** charging time, **b** infrastructure cost per station, and **c** grid load impact

to traditional gasoline refueling [3, 6]. BSSs maintain an inventory of charged batteries and charge them during off-peak hours, facilitating grid-friendly operations and supporting renewable integration [9, 14]. This approach also enables centralized battery ownership and lifecycle management. Operators can monitor battery health, optimize charging strategies, and extend battery lifespan [11]. By decoupling vehicle and battery ownership, users avoid battery degradation concerns and benefit from reduced upfront EV costs [1]. A comparison of charging methods in terms of time, infrastructure cost, and grid impact is illustrated in Fig. 2, providing a visual summary of key differences.

2.1.3 Comparative Analysis

- **Time Efficiency**: Battery swapping provides instant energy replenishment, surpassing all plug-in methods.
- **Infrastructure Investment**: Swapping requires higher upfront investment and standardization but enables high-throughput operation.
- **Grid Impact**: Off-peak battery charging allows BSSs to alleviate grid congestion, unlike fast charging during peak hours.
- **User Convenience**: Swapping addresses range anxiety and supports seamless energy access for high-utilization fleets.

The consolidated benefits of BSS are illustrated in Fig. 3.

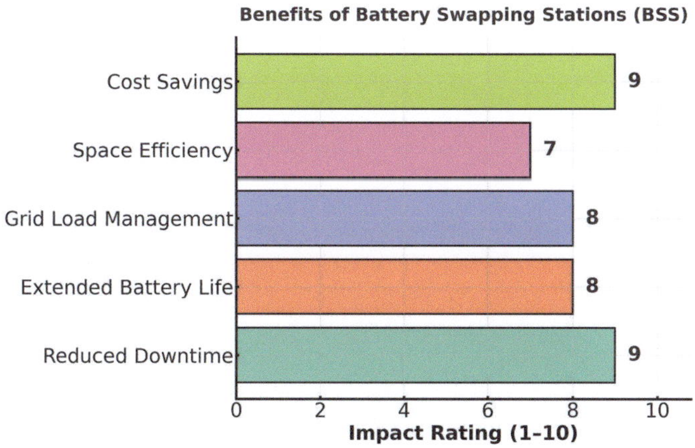

Fig. 3 Summary of key benefits of BSS

2.2 *Benefits of BSS*

Battery swapping stations present several benefits over conventional charging:

- **Reduced Downtime**: Battery swaps are completed in minutes, ideal for fleet efficiency.
- **Extended Battery Life**: Centralized, optimized charging improves longevity.
- **Grid Load Management**: Charging during off-peak hours supports load shifting and grid balancing [9].
- **Space Efficiency**: Swapping infrastructure is compact and suitable for dense urban areas [15].
- **Lower Upfront Cost**: Battery-as-a-Service (BaaS) models reduce EV acquisition costs and separate battery ownership [1].

2.3 *Key Players and Existing Deployments*

Several companies around the world are actively developing and deploying battery swapping infrastructure through commercial rollouts and strategic partnerships. Table 2 summarizes some of the key players, their geographical focus, and notable deployment highlights, spanning applications from two-wheelers to modular fleet systems.

Table 2 Key players and global battery swapping deployments

Company	Country	Deployment highlights
NIO	China	1,300+ stations, millions of swaps, battery upgrades and BaaS
Gogoro	Taiwan	Leading two-wheeler network, urban coverage, ease of use
Ample	USA	Modular, vehicle-agnostic, integrated into delivery and ride-hailing fleets
SUN Mobility	India	Focused on 2W/3W, app-based interface and multi-OEM support
CATL	China	Large-scale expansion, OEM collaborations on battery standards

3 Integration of Renewable Energy in BSS

3.1 Renewable Energy Sources for BSS

Integrating renewable energy into BSS enhances sustainability, lowers greenhouse gas emissions, and improves energy autonomy. The primary renewable sources suitable for BSS include solar photovoltaic (PV), wind energy, and hybrid renewable energy systems. Utilizing these sources not only reduces operational costs but also aligns with global decarbonization targets [14, 15, 21].

3.1.1 Solar Photovoltaic (PV)

Solar PV systems are among the most practical and scalable renewable options for powering BSS. By converting solar radiation into electrical energy, PV panels can directly supply electricity for charging swappable batteries during daylight hours. Installing PV arrays on rooftops, canopies, or adjacent open land allows BSS operators to capitalize on localized generation, reducing dependence on grid electricity. Additionally, government incentives and the declining cost of PV modules enhance the economic viability of solar-powered BSS. Case studies have shown that solar-powered BSSs significantly reduce energy bills and operational emissions, especially in regions with high solar insolation. Furthermore, PV systems are modular and scalable, allowing station operators to incrementally increase capacity based on user demand and local site conditions [14].

3.1.2 Wind Energy

Wind energy, while more site-specific than solar, presents a valuable complement to PV in locations with favorable wind resources. Small- and medium-sized wind turbines can be deployed near BSS to provide electricity for battery charging, particularly during non-daylight hours. This is particularly beneficial in coastal areas or highlands where wind speeds remain consistent throughout the day and night. Wind energy contributes to the overall energy balance of hybrid systems, compensating for periods when solar generation is low (e.g., during cloudy weather or nighttime). Integration of wind power into BSS helps flatten the power curve and reduces reliance on the grid. In addition, wind energy integration supports off-grid or partially grid-connected BSS deployments in remote or underdeveloped regions [21].

3.1.3 Hybrid Systems

Hybrid renewable systems combine solar, wind, and occasionally other resources like hydrogen fuel cells or biogas-based generation. These systems offer superior reliability by mitigating the intermittency associated with single-source generation. For instance, hybrid PV-wind configurations can supply continuous power by leveraging complementary generation profiles—solar during daytime and wind during night hours or cloudy days. Hybrid systems are often supported by advanced control systems that optimize energy dispatch, maximize renewable penetration, and manage battery storage. Simulation-based studies demonstrate that hybridized renewable configurations for BSS reduce both levelized cost of energy (LCOE) and carbon footprint, making them a cornerstone of future-ready EV infrastructure [11, 23].

3.2 Energy Storage Systems and Grid Interfacing

Energy Storage Systems (ESSs) are indispensable in renewable-integrated BSS architectures. They act as a buffer between variable renewable generation and dynamic load profiles, ensuring uninterrupted battery charging even when renewable supply fluctuates. Lithium-ion batteries are the most commonly used ESS due to their high energy density, long cycle life, and fast response time. ESS enables the station to store excess solar or wind energy generated during peak production periods and discharge it during demand spikes or renewable shortfalls. In grid-connected scenarios, ESS facilitates time-shifting of grid energy consumption—drawing power during off-peak hours and feeding stored energy back to the grid during peak demand, participating in demand response (DR) markets [9]. Moreover, advanced ESS setups often include energy management controllers, charge/discharge optimizers, and safety subsystems for temperature control, overvoltage protection, and fault handling. Integration with Building Energy Management Systems (BEMS) further enhances grid interoperability and supports applications like black-start capability, voltage smoothing, and renewable curtailment mitigation [21].

3.3 Power Electronics and Conversion Interfaces

Power electronics serve as the foundation for seamless energy exchange between renewable generators, ESS, the grid, and swappable EV batteries. The most critical power electronic components in a BSS include the following:

- **DC–DC converters**: Regulate the voltage between renewable inputs and the ESS or batteries, optimizing charge levels and ensuring safety.
- **Inverters**: Convert DC from solar panels or battery banks into AC compatible with grid infrastructure and auxiliary loads.
- **Rectifiers**: Convert AC grid power to DC for direct battery charging, especially in grid-assisted BSS.
- **Charge controllers and MPPT units**: Maximize power extraction from PV arrays and ensure battery charging stays within thermal/electrical limits.

Recent research emphasizes the use of bidirectional inverters and modular multilevel converters (MMC) in BSS to support both charging and discharging cycles in grid-interactive modes (G2B and B2G). These technologies improve energy transfer efficiency, enable real-time fault detection, and reduce harmonic distortion [11, 15].

3.4 Smart Energy Management Systems (EMS)

A Smart EMS is the digital intelligence of a modern BSS. It coordinates the interaction among renewable energy generation, storage units, grid interfacing, and vehicle battery demand in real time. EMS utilizes IoT-enabled sensors, predictive control algorithms, and AI-based demand forecasting to manage power flow and prioritize energy dispatch.

Key EMS functions include the following:

- Dynamic load forecasting and renewable generation prediction.
- Real-time battery availability tracking.
- G2V/V2G operation scheduling.
- Renewable-first charging optimization.
- Demand response event participation.

By balancing supply and demand in milliseconds, EMS improves operational efficiency, minimizes curtailment losses, and extends system lifespan. Moreover, cloud-enabled EMS platforms support remote diagnostics, firmware updates, and fleet-wide energy coordination across multiple BSS sites [10, 23].

3.5 Grid-to-Battery (G2B) and Battery-to-Grid (B2G) Operations

G2B and B2G operations transform BSS from energy consumers into active grid participants. **G2B** involves charging batteries from the grid during periods of low electricity demand and low tariffs. This allows operators to reduce electricity costs and alleviate stress on the grid during high-load periods. G2B scheduling is typically governed by time-of-use pricing and automated EMS algorithms. **B2G** allows energy stored in EV batteries or ESS to be discharged back into the grid, particularly during peak demand events or emergencies. This enables BSS to offer ancillary services such as:

- Frequency and voltage regulation.
- Spinning reserve support.
- Peak shaving and load leveling.

As a result, BSSs equipped with B2G capability serve as distributed energy resources (DERs), strengthening grid resilience while opening new revenue streams for operators. Emerging policies in countries like China, USA, and India are accelerating B2G adoption through regulatory support and financial incentives [1, 10].

4 System Architecture and Operational Framework

4.1 Components of a Renewable-Based BSS

A renewable-based BSS consists of several critical components, each contributing to energy sustainability, operational efficiency, and intelligent management. The overall system architecture is depicted in Fig. 4, which illustrates the interactions among renewable sources, energy storage, power conversion, and smart control systems. At the core, solar PV panels and wind turbines serve as the primary renewable energy sources, generating clean electricity on-site. These sources are interfaced with power electronic converters, including DC–DC converters, inverters, and maximum power point tracking (MPPT) controllers, which manage voltage, frequency, and current compatibility between the generation side, storage, and battery systems [14, 15].

ESS provides a buffer that absorbs surplus energy during high-generation periods and releases stored energy when renewable availability drops. ESS helps balance energy supply with battery swapping demand, offering load-shifting capabilities and reducing reliance on grid imports [21]. The station includes a battery storage and management unit where charged and depleted swappable batteries are organized, monitored, and thermally managed. These facilities are often integrated with robotic swapping mechanisms that enable fully automated battery replacement, minimizing downtime and maximizing operational throughput [1]. A centralized EMS orchestrates the flow of energy across the system-tracking state of charge (SOC), forecasting

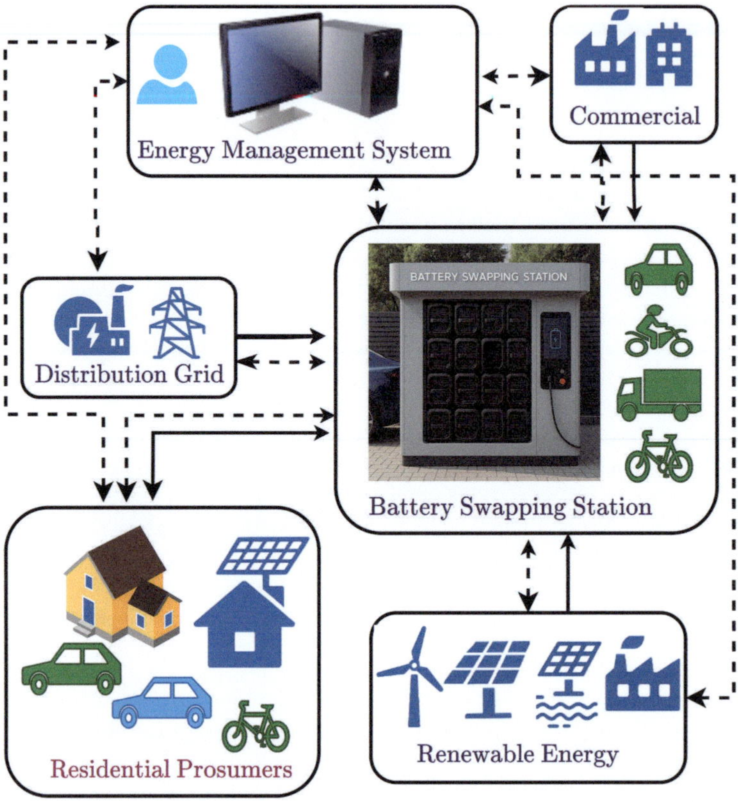

Fig. 4 System-level architecture of a renewable-powered BSS integrated with energy storage, grid interface, and EMS

renewable availability, predicting load demand, and executing cost-optimized control strategies. EMS also ensures system safety, logs diagnostics, and communicates with external grid interfaces for G2B and B2G operations [10].

4.2 Scheduling Techniques (E.g., Day-Ahead Scheduling)

Effective scheduling is essential for aligning energy availability with demand at battery swapping stations. Day-ahead scheduling is commonly employed in renewable-integrated BSS to preplan the charging, discharging, and battery swapping cycles based on predicted renewable generation and station load for the upcoming day. By forecasting solar irradiance, wind availability, and anticipated EV traffic, the EMS can determine when to charge batteries using renewable energy, when to draw from the grid, and when to store or release energy through the ESS. Such planning maximizes clean energy usage and minimizes grid draw during peak tariff hours [9, 14]. Moreover, day-ahead scheduling facilitates demand response participation by allowing BSS to curtail or defer loads during high-demand grid periods. It enables time-of-use optimization and economic dispatch of energy resources, especially in regions with variable electricity pricing schemes [23].

4.3 Load Forecasting and Demand Estimation

Accurate load forecasting is vital to the reliable operation of a battery swapping station. Forecasting involves the use of historical data, user behavior patterns, and external factors to anticipate future demand for charged batteries and electricity. Key factors influencing demand forecasts include time-of-day effects, weekly or seasonal trends, weather conditions (which affect renewable output), and special events that could lead to abnormal spikes in EV usage. Predictive algorithms, often based on machine learning or probabilistic modeling, are employed by EMS platforms to produce short-term and long-term demand profiles [15, 21]. Real-time correction mechanisms are also used to adjust forecasts on-the-fly, enabling the system to respond adaptively to unexpected changes. This proactive forecasting ensures the availability of sufficient charged batteries while avoiding overcharging and energy wastage.

4.4 Optimization Models for Energy and Cost Management

Optimization models are applied in renewable-based BSS to minimize energy costs, improve system efficiency, and enhance environmental sustainability. These models are embedded within EMS platforms and support decision-making related to energy dispatch, charging schedules, and grid interaction. Energy dispatch optimization aims to identify the most cost-effective energy sources at any given time. For instance, if solar energy is abundant, batteries are charged from PV, while grid input is minimized. Conversely, during low renewable output, batteries may be charged from ESS or the grid based on cost–benefit evaluation [11]. Multi-objective optimization frameworks consider trade-offs among operational cost, battery degradation, carbon emissions, and energy efficiency. These models often employ techniques such as linear programming, dynamic programming, or evolutionary algorithms to generate optimal control strategies under dynamic constraints [1, 10]. Overall, such optimization ensures that BSS operates in an economically viable and environmentally responsible manner, supporting the wider goal of sustainable electric mobility.

5 Global Initiatives and Case Studies

5.1 Notable International Projects and Pilots

Battery swapping has emerged as a strategic solution to address the limitations of conventional EV charging infrastructure, particularly in urban and high-utilization scenarios. Countries around the world are piloting and scaling battery swapping infrastructure to evaluate its feasibility, user acceptance, and grid integration potential. In the United States, the startup Ample has pioneered modular battery swapping technology designed for compatibility with a wide range of EV models. Their systems are currently deployed in pilot programs targeting ride-hailing and delivery fleets, aiming to reduce vehicle downtime and operational cost. Ample's approach leverages standardized battery packs and robotic swapping units to enable under-five-minute swaps, aligning with the needs of high-throughput fleet services.

In Taiwan, Gogoro has achieved remarkable success with its two-wheeler battery swapping ecosystem. As of early 2021, the company had deployed over 1,900 GoStations and supported more than 760,000 battery packs in circulation. Gogoro's network accounts for a large portion of Taiwan's electric scooter market and demonstrates the effectiveness of user-centric infrastructure combined with strong policy support and strategic partnerships. Gogoro has also extended its operations to Indonesia, where it launched a pilot project in Jakarta in collaboration with Gojek. This initiative deployed several hundred electric scooters and GoStations to promote sustainable last-mile transportation in congested urban areas. Germany has also expressed growing interest in battery swapping, with government-funded projects involving companies like Bosch and research institutions. These initiatives are exploring the standardization and technical viability of battery swapping in the context of European mobility ecosystems, where skepticism and infrastructure inertia remain challenges to adoption.

5.2 Success Stories from India and China

China leads the world in battery swapping deployment and innovation, having established a robust policy and industrial framework to support its growth. NIO, one of the country's leading EV manufacturers, has constructed over 2,300 battery swapping stations nationwide as of 2023. These stations allow users to replace a depleted battery with a fully charged one in under five minutes and even upgrade their battery capacity without changing vehicles. NIO's success showcases how vertically integrated design, software control, and infrastructure scale can converge to deliver seamless EV user experiences [1].

Additionally, CATL, the world's largest battery manufacturer, has partnered with the energy giant Sinopec to construct 10,000 battery swap stations across China. This partnership underscores the country's commitment to making battery swapping viable not only for passenger vehicles but also for commercial and heavy-duty trucks. The collaboration aligns with broader national goals to enhance energy efficiency, reduce emissions, and expand zero-emission transportation networks. In India, battery swapping is being explored as a practical solution for electrifying two- and three-wheelers, which dominate urban transport. SUN Mobility is a key player offering interoperable battery swapping solutions through its Quick Interchange Stations (QISs). The company has established partnerships with multiple OEMs and fleet operators to integrate swappable battery systems in electric autorickshaws and delivery scooters. Another notable development is Gogoro's entry into the Indian market through a collaboration with Zypp Electric. The pilot, launched in Delhi in 2022, focuses on last-mile delivery operations and aims to replicate the success of Gogoro's Taiwan model by adapting it to India's unique urban logistics demands.

5.3 Comparative Performance Analysis

A comparative evaluation of initiatives in China and India reveals both convergence and divergence in deployment strategies, shaped by local transportation dynamics, government policy, and industrial capability. China's battery swapping efforts are characterized by scale, technological sophistication, and integration with national EV and grid strategies. The emphasis is on passenger vehicles, fleet logistics, and increasingly, heavy-duty trucks, supported by large-scale infrastructure investment and standardized vehicle platforms. In contrast, India's battery swapping model focuses predominantly on the two- and three-wheeler segment, reflecting the mobility needs of its urban population. The Indian government is in the process of drafting supportive policy frameworks to address battery standardization, interoperability, and business models that enable private-sector participation. While still in early phases compared to China, India's focus on affordability, compact station design, and decentralized mobility gives it a distinctive edge in scalable, bottom-up deployment.

Both countries recognize the need to decouple vehicle ownership from battery ownership, a paradigm that allows for lower EV acquisition costs and centralized battery management. China has already demonstrated that large-scale manufacturer-driven swapping systems can achieve high user satisfaction and grid benefits. Meanwhile, India is carving a path based on modular infrastructure and fleet electrification, particularly for gig-economy and logistics operators.

5.4 Lessons Learned and Best Practices

Global experiences with battery swapping yield several actionable lessons. First, standardization of battery formats and connectors is critical to ensure cross-compatibility across vehicle models and station networks. Countries like China have succeeded in part because major automakers aligned early with standardized battery and chassis designs, enabling rapid infrastructure rollout. Second, public–private partnerships have proven essential for scaling deployment. In all successful examples, such as those led by NIO, Gogoro, or SUN Mobility, collaboration between governments, utilities, and private technology providers has facilitated land access, grid connectivity, and regulatory approvals. Third, user-centric design is essential. Swapping systems must be quick, reliable, and integrated with mobile apps or cloud platforms for real-time battery tracking and station availability. The convenience of a five-minute battery swap, when coupled with digital services, has significantly enhanced adoption in regions like Taiwan and parts of China.

Finally, coupling battery swapping infrastructure with renewable energy sources significantly improves long-term sustainability. Integrating solar PV or wind systems with storage and EMS at the station level reduces grid dependency and supports clean mobility goals. Studies in multiple countries, including hybrid PV-based BSS pilots in India and China, affirm the economic and environmental benefits of such integration [14, 15].

6 Techno-Economic and Environmental Analysis

6.1 Cost–Benefit Analysis of Renewable-Based BSS

Implementing BSS powered by renewable energy introduces a multi-dimensional cost–benefit dynamic that blends capital-intensive infrastructure with long-term savings and environmental benefits. The initial investment required for a renewable-based BSS includes the cost of installing solar PV arrays, wind turbines, ESS, and robotic swapping mechanisms. These expenses are substantial and often represent a barrier to entry, particularly for decentralized or small-scale operators. However, in regions with high solar irradiance or wind availability, the investment can be offset over time by reduced electricity procurement costs and peak demand charges [14, 21].

Operational expenses vary based on the technology stack and business model. On-site renewable energy significantly lowers recurring energy costs, especially when paired with intelligent energy management systems that optimize dispatch. However, maintenance of renewable installations, battery diagnostics, and swapping system upkeep add to recurring expenditures. Comprehensive techno-economic models are necessary to weigh these dynamics and to assess lifecycle profitability, especially under different utilization and incentive scenarios [23].

On the revenue side, BSS can generate income through direct battery swapping services, selling excess renewable electricity back to the grid, and participating in ancillary markets such as frequency response or peak shaving. Some operators also explore Battery-as-a-Service (BaaS) or battery leasing models, which distribute costs across subscription users and improve cash flow predictability. The financial viability of such models is strongly tied to local policy frameworks, tariff design, and mobility demand patterns. Additionally, the break-even period for a renewable-based BSS depends on multiple factors, including capital structure, energy prices, and government incentives. In some regions, policy interventions and carbon credits can shorten the payback period significantly [1].

6.2 Impact on Grid Load and Stability

Battery swapping stations have a notable influence on grid stability and demand management, especially when integrated with variable renewable energy sources. When coordinated effectively through an EMS, BSS can perform load-shifting operations—charging batteries during periods of low demand or high renewable generation, and deferring loads during peak grid stress periods. This strategic scheduling alleviates strain on the distribution network, optimizes transformer loading, and reduces the need for costly grid upgrades [9, 23]. Moreover, renewable-based BSS can participate in grid-support services. With fast response times and bidirectional inverters, energy stored in ESS or idle EV batteries can be injected back into the grid to support frequency regulation, voltage control, and contingency reserves. These capabilities are increasingly valuable in power systems with high penetration of intermittent renewables like wind and solar, which introduce volatility into supply–demand balancing. However, the integration of renewable sources into BSS also presents operational challenges. The variable nature of solar and wind energy demands robust forecasting, energy buffering, and real-time optimization to avoid fluctuations that could destabilize local grid nodes. Without adequate planning, high-density BSS deployment may also increase local peak load conditions during unexpected high-demand periods. To counteract this, advanced EMS frameworks employing reinforcement learning and demand prediction algorithms are being explored to dynamically adapt load patterns and maintain grid balance [11, 21].

6.3 Environmental Benefits and Carbon Footprint Reduction

The environmental benefits of renewable-powered BSS are a major driver of their global adoption. By utilizing clean energy sources for charging EV batteries, BSSs substantially reduce indirect emissions associated with electric transportation. In regions where the grid is still dependent on coal or gas, on-site renewables help decouple EV operations from fossil-fuel-based electricity, improving the carbon intensity of mobility services [15]. Quantitative studies have shown that widespread deployment of electric two-wheelers using BSS could result in significant greenhouse gas (GHG) reductions. For example, in high-density urban centers in Asia, electric motorcycles with battery swapping systems have demonstrated CO_2 savings in the range of 2.6–3.0 million tons annually, compared to gasoline-powered counterparts [10]. Beyond operational emissions, BSSs also promote better battery lifecycle management. Since batteries are centrally owned and monitored, their usage can be optimized, and end-of-life processes such as recycling and second-life applications (e.g., stationary storage) can be streamlined. This minimizes resource waste and environmental risks associated with decentralized battery disposal. Furthermore, integrating energy-efficient power conversion systems, smart cooling, and passive design strategies within BSS facilities enhances sustainability performance. Coupled with government-backed clean mobility programs, renewable-based BSSs are well positioned to advance national and international decarbonization goals.

6.4 Business Models and Market Viability

The economic success of renewable-based BSS is heavily contingent on the adoption of viable business models and enabling policy environments. A wide range of ownership structures have emerged, including utility-led models, OEM-integrated BSS ecosystems, and service-provider-based frameworks. Battery leasing models, in particular, lower the upfront cost of EVs and increase affordability, making electric mobility accessible to broader user segments. Collaborations and partnerships are critical for the expansion of BSS infrastructure. Strategic alliances between energy utilities, mobility service providers, and battery manufacturers allow for shared capital investment, technical co-development, and joint branding. Such synergies improve supply chain stability and accelerate market penetration. Supportive regulation is equally essential. Standardization of battery packs and connectors is required to enable interoperability across vehicle brands and swapping stations. Countries like China have made significant progress in this area, while others are in the early stages of policy drafting. Incentives such as capex subsidies, land access for stations, feed-in tariffs for renewable excess energy, and tax credits for BSS operators can dramatically improve project bankability and investor confidence. Ultimately, market viability depends on creating scalable, user-friendly systems that deliver value to all stakeholders-fleet operators, individual users, energy providers, and governments. With the right mix of technology, regulation, and market design, renewable-based battery swapping stations can become a cornerstone of clean urban mobility ecosystems.

7 Challenges and Limitations

Despite the promising potential of renewable-based BSS, several technical, economic, and regulatory barriers must be addressed to ensure sustainable and scalable deployment. These challenges span across energy supply reliability, system interoperability, user adoption, and policy development.

7.1 Renewable Energy Intermittency and Storage

The intermittent nature of renewable energy sources—particularly solar and wind—poses a fundamental operational challenge for BSS. Solar energy production depends on sunlight availability, which fluctuates daily and seasonally, while wind energy output is influenced by changing wind speeds and weather patterns. These fluctuations result in inconsistent power generation, which may not always align with the real-time energy demand for charging batteries at swapping stations [15]. To mitigate this, ESSs are required to buffer the supply–demand mismatch and ensure reliable energy availability during low-generation periods. However, deploying ESS involves additional investment in hardware, space, and maintenance, which raises the overall capital expenditure of renewable-integrated BSS infrastructure [21]. Moreover, the performance and cost of ESS are influenced by battery chemistry, lifecycle degradation, and environmental conditions, all of which demand robust control strategies and long-term system planning. Advanced energy management systems with predictive capabilities are needed to optimize ESS utilization, forecast renewable availability, and coordinate with grid imports when necessary [23].

7.2 Standardization and Interoperability

The lack of standardization in battery formats, communication protocols, and mechanical interfaces presents a major barrier to the universal deployment of BSS. EV manufacturers currently design proprietary battery packs optimized for specific vehicle models. As a result, the development of swapping stations that can serve diverse vehicle types is limited, and infrastructure interoperability suffers [1]. Without standardized battery dimensions, connectors, thermal interfaces, and communication protocols, every station must either be designed for a specific brand or accommodate multiple complex configurations—both of which increase engineering and operational costs. Standardization not only enables scale economies but also enhances user convenience and service reliability. On the regulatory front, the absence of unified international or national guidelines governing battery swapping processes—such as safety compliance, data exchange standards, and battery testing protocols—further hinders large-scale deployment. The establishment of global or region-specific standards by regulatory bodies is essential to facilitate interoperability, ensure safety, and encourage industry collaboration [11].

7.3 Battery Ownership and Swapping Models

The question of battery ownership is central to the economic and operational design of battery swapping systems. In traditional EV models, the battery constitutes a significant portion of the vehicle's cost and is owned by the consumer. However, in a swapping paradigm, ownership is often transferred to the station operator or a third-party energy service provider. This shift raises questions related to liability, insurance, lifecycle management, and residual value [10]. Operators must ensure uniform battery quality, implement robust battery tracking systems, and establish clear service-level agreements (SLAs) for battery health and performance. From the user's perspective, concerns may arise regarding receiving an older or degraded battery in exchange for a newer one. These concerns can hinder consumer confidence and adoption, especially in markets where EV buyers are unfamiliar with battery-as-a-service (BaaS) models. To alleviate these concerns, some operators employ centralized battery health monitoring, smart diagnostics, and usage-based billing to ensure fairness and transparency. Nevertheless, successful implementation requires building trust through clear communication, policy clarity, and flexible service contracts.

7.4 Regulatory and Policy Barriers

Policy and regulatory support plays a crucial role in the growth and viability of renewable-powered BSS. In many countries, the absence of targeted government incentives, financial subsidies, or supportive grid tariffs creates unfavorable investment environments. Unlike direct EV charging infrastructure, battery swapping may fall into policy blind spots or be treated under outdated regulatory classifications, limiting access to land, grid interconnection, and renewable energy subsidies [9]. Moreover, safety regulations governing battery handling, thermal runaway protection, and fire suppression systems are often lacking or inconsistently applied across jurisdictions. Establishing standardized codes and enforcement mechanisms for battery storage, transport, and swapping procedures is vital to ensure user safety and operational reliability [23]. Another critical policy challenge is aligning battery swapping frameworks with larger energy and mobility objectives. Incentives such as capex support for swapping stations, tax relief for operators, feed-in-tariffs for renewable excess energy, and standardization mandates can significantly boost adoption. In regions where such policies have been introduced—such as in China or Taiwan—battery swapping has scaled rapidly and sustainably. For emerging markets, proactive regulatory design is essential to replicate this success and integrate battery swapping into national clean mobility strategies.

Figure 5 highlights the key technical, regulatory, and operational challenges that constrain the scalability of renewable-based BSS. Issues such as renewable intermittency, lack of standardization, regulatory uncertainty, and consumer trust concerns collectively affect system reliability, interoperability, and economic viability. Addressing these multifaceted limitations requires an integrated and coordinated approach. A structured summary of these challenges and their corresponding mitigation strategies is provided in Table 3.

Fig. 5 Key challenges affecting the scalability and reliability of renewable-based BSS

8 Future Research Directions and Opportunities

8.1 Advanced Optimization and AI-Based Control

The incorporation of artificial intelligence (AI) and machine learning techniques presents significant opportunities to optimize the performance of BSS. Predictive analytics powered by AI can forecast both energy supply-based on solar irradiance or wind patterns and energy demand based on historical usage, mobility trends, and grid tariffs. This enables dynamic scheduling of battery charging and swapping activities, improving the alignment between renewable availability and load demand [11]. Furthermore, machine learning algorithms can analyze real-time station data to enhance inventory management, predict battery degradation, optimize energy dispatch, and reduce downtime. Such systems can autonomously adjust operational parameters in response to fluctuating energy prices or demand surges, thereby lowering operational costs and improving service reliability [23].

Table 3 Key challenges in renewable-based BSS and potential mitigation strategies

Challenge	Implication	Potential mitigation
Renewable energy intermittency	Fluctuating energy supply affects battery charging availability	Integration of ESS, predictive EMS, hybrid renewable systems
High capital investment	Increased upfront costs for renewable infrastructure and automation	Financial subsidies, public–private partnerships, long-term energy savings
Lack of standardization	Incompatibility across vehicle OEMs and batteries, reduced interoperability	Development of universal battery standards, regulatory frameworks
Unclear battery ownership models	Confusion over liability, maintenance, and cost allocation	Battery leasing, centralized health monitoring, usage-based pricing
Consumer trust	Reluctance to swap unknown batteries due to quality concerns	Real-time battery diagnostics, transparency apps, warranty assurance
Policy and regulatory gaps	Limited access to incentives or unclear guidelines for swapping stations	Targeted BSS policies, safety codes, integration into EV roadmaps
Grid integration complexity	Risk of instability due to uncoordinated charging patterns	Smart EMS, time-of-use optimization, demand response participation

8.2 Integration with Smart Grids and V2G Systems

Another promising research avenue lies in the integration of BSS with smart grid infrastructure and vehicle-to-grid (V2G) capabilities. Smart grids equipped with advanced metering infrastructure (AMI) and two-way communication protocols allow BSS to participate in grid-interactive operations. Bidirectional energy flow enables EV batteries to act as distributed energy storage units, feeding electricity back into the grid during peak demand periods and absorbing excess generation during troughs [9]. In this context, BSS can play a vital role in demand response programs, responding to dynamic price signals or frequency regulation events to enhance grid flexibility. Coordinated EMS and V2G functionalities could transform BSS into energy prosumers—consuming, storing, and redistributing electricity as per system needs [10].

8.3 Blockchain for Energy Transactions

Blockchain technology offers a novel solution for transparent, secure, and decentralized energy management in BSS ecosystems. By creating tamper-proof, real-time records of energy generation, usage, and swaps, blockchain ensures trust among various stakeholders, including energy providers, BSS operators, and EV users. Smart contracts can be used to automate billing, enforce service-level agreements, and facilitate micro-transactions with minimal overhead [15]. Furthermore, decentralized energy trading platforms built on blockchain allow peer-to-peer energy exchanges, wherein BSS units can buy, sell, or trade electricity based on demand forecasts, renewable availability, or grid pricing—creating a more efficient and resilient local energy market [23].

8.4 Scalability and Urban Deployment Models

Scaling BSS infrastructure in dense urban environments presents logistical and spatial challenges that require innovative deployment models. Future research must focus on space optimization techniques, including modular station layouts, vertical battery storage, and integration with existing urban infrastructure such as parking lots and transit hubs. Urban planning tools can assist in identifying high-demand zones based on traffic flow, ride-sharing patterns, and population density, thereby guiding optimal site placement [14]. Moreover, digital twin models of urban areas can simulate energy demand, vehicle flow, and renewable generation potential to evaluate the impact of BSS deployment on traffic congestion, air quality, and urban energy resilience. Incorporating these insights into policy and planning can accelerate large-scale efficient rollout of BSS in smart cities.

9 Conclusion

9.1 Summary of Key Insights

This chapter has provided a comprehensive overview of renewable energy-based BSS as a sustainable and scalable alternative to conventional EV charging infrastructure. Battery swapping addresses critical barriers such as long charging times and range anxiety, offering fast, user-friendly energy replenishment. The integration of renewable sources—primarily solar and wind—into BSS significantly reduces the carbon footprint of EV infrastructure while enhancing energy resilience and grid independence. Efficient operation of BSS relies on the deployment of smart EMS, energy storage systems, and advanced scheduling strategies, including day-ahead forecasting. These components ensure optimal coordination between energy generation, storage, and demand. International case studies, particularly from China and India, have demonstrated the real-world viability of BSS, showcasing diverse business models and technical implementations. However, widespread adoption still faces challenges related to battery standardization, regulatory frameworks, and high capital investment.

9.2 Strategic Recommendations

To accelerate the adoption and scale-up of renewable-powered BSS, coordinated efforts across policy, technology, and industry are essential. Governments should introduce supportive measures such as financial incentives, tax exemptions, and capital subsidies to attract investment in both infrastructure and renewable integration. Establishing industry-wide standards for battery formats, swapping protocols, and safety compliance is critical for ensuring interoperability and simplifying deployment. Public–private partnerships can play a catalytic role by leveraging the expertise and resources of government agencies, energy providers, and EV manufacturers. Such collaborations are key to developing reliable, scalable, and cost-effective BSS ecosystems. Additionally, sustained research and innovation are required in areas such as AI-based energy optimization, next-generation battery technologies, and space-efficient urban BSS models to further enhance system performance and user experience.

9.3 Vision for the Future of Green Battery Swapping Ecosystems

Looking ahead, the convergence of renewable energy, intelligent control systems, and interoperable swapping infrastructure presents a transformative pathway for electric mobility. Battery swapping can serve as a cornerstone of smart, decentralized energy systems—enabling net-zero emissions, grid flexibility, and mass EV adoption across public and private transportation sectors. Future BSS ecosystems will not only enable clean and rapid vehicle electrification but also support broader sustainability objectives by functioning as distributed energy hubs. When embedded within urban mobility networks, powered by clean energy, and managed by AI-driven platforms, BSS can reshape the energy-transport nexus into a system that is efficient, environmentally sound, and economically inclusive. As a bridge between mobility and energy innovation, BSS offers a scalable solution in the global transition toward sustainable urban transport.

References

1. Renga D, Meo M (2025) Modeling battery swapping stations for sustainable urban mobility. Sustain Energy Grids Netw 41:101592
2. Charoen-amornkitt P, Nantasaksiri K, Ruangjirakit K, Laoonual Y (2024) Energy consumption and carbon emission assessment of battery swapping systems for electric motorcycles. Heliyon 10(1):e18677
3. Adegbohun F, von Jouanne A, Lee KY (2019) Autonomous battery swapping system and methodologies of electric vehicles. Energies 12(4):667
4. Kundu S, Giri AK, Kadiyan S, Singh S (2025) Power management and adaptive control method for a standalone distributed generation system under fault, dc-offset, and dynamic environments. Int J Ambient Energy 46(1):1–14
5. Singh N, Kundu S, Verma V, Nangru Y (2025) Effective energy efficiency methods for distribution transformers. In: 2025 First international conference on advances in computer science, electrical, electronics, and communication technologies (CE2CT). IEEE, pp 579–584
6. Revankar SR, Kalkhambkar VN (2021) Grid integration of battery swapping station: a review. J Energy Storage 41:102937
7. Kundu S, Giri A, Singh M, Singh S (2025) Effective multiport battery charging facility, operation and control in hybrid isolated microgrid system. In: Giri A, Singh M (eds) Electric vehicle charging infrastructures and its challenges. Studies in infrastructure and control. Springer, Singapore
8. Kundu S, Giri AK, Kadiyan S (2025) An adaptive mixed-step size normalized least means fourth control approach for stand-alone power generation system considering dynamic conditions. IEEE J Emerg Select Top Power Electron 13:2422–2429
9. Renga D, Spoturno F, Meo M (2024) Reinforcement learning for charging scheduling in a renewable powered battery swapping station. IEEE Trans Veh Technol 73(10):14382–14398
10. Chien Y-H, Hsieh I-YL (2025) Enhancing urban energy resilience: vehicle-to-grid (v2g) strategies within electric scooter battery swapping ecosystems. Sustain Cities Soc 118:106004
11. Yang J, Yu F, Ma K, Yang B, Yue Z (2024) Optimal scheduling of electric-hydrogen integrated charging station for new energy vehicles. Renew Energy 224:120224
12. Kundu S, Singh M, Giri A, Kumar V (2024) Robust and fast control approach for islanded microgrid system and ev charging station applications. Electric Eng
13. Kundu S, Singh M, Giri A (2024) Spv-wind-bes-based islanded electrical supply system for remote applications with power quality enhancement. Electr Eng 106:279–294
14. Ban M, Yu J, Shahidehpour M, Guo D (2019) Optimal sizing of pv and battery-based energy storage in an off-grid nanogrid supplying batteries to a battery swapping station. J Modern Power Syst Clean Energy 7(2):309–320
15. Zhang C, Lu X, Chen S, Shi M, Sun Y, Wang S, Zhang S, Fang Y, Zhang N, Foley AM, He K (2024) Synergies of variable renewable energy and electric vehicle battery swapping stations: case study for Beijing. eTransportation 22:100363
16. Kundu S, Singh M, Giri AK (2024) Adaptive control approach-based isolated microgrid system with alleviating power quality problems. Electric Power Compon Syst 52(7):1219–1234
17. Kundu S, Giri A, Singh M, Kadiyan S (2024) Synchronization and control of wecs-spv-bss-based distributed generation system using iccf-pll control approach. Electric Power Syst Res 226:109919
18. Mohanty P, Pradhan R, Jena P, Padhy N (2025) Powering the future: an in-depth exploration of global electric vehicle charging infrastructure. In: Giri A, Singh M (eds) Electric vehicle charging infrastructures and its challenges. Studies in infrastructure and control. Springer, Singapore
19. Kocer MC, Onen A, Ustun TS, Albayrak S (2022) Optimization of multiple battery swapping stations with mobile support for ancillary services. Front Energy Res 10:945453
20. Yap KY, Chin HH, Klemes JJ (2025) Solar energy-powered battery electric vehicle charging stations: current development and future prospect review. Renew Sustain Energy Rev 173:113365

21. Nyamayoka LT-E et al (2025) Techno-economic feasibility and optimal design approach of grid-connected hybrid power generation systems for electric vehicle battery swapping station. Energies 18(5):1208
22. Mohanty P, Pradhan R, Jena P, Padhy N (2025) Electric vehicles: exploring types, benefits, challenges, policies, and smart charging innovation. In: Giri A, Singh M (eds) Electric vehicle charging infrastructures and its challenges. Studies in infrastructure and control. Springer, Singapore
23. Xu X et al (2022) Robust energy management for an on-grid hybrid hydrogen refueling and battery swapping station based on renewable energy. J Clean Prod 331:129954

Chapter 11
Case Study for Integration of Renewable Sources to Loharu Micro Grid to Reduce the Expenses and Emissions Using HOMER Pro

Surender Singh

Abstract In remote rural parts of developing countries, access to electricity is still a major problem since grid expansion is expensive and challenging to operate. The hybrid renewable energy system proposed in this investigation is based on an extensive assessment of the domestic load of 437 houses and significant facilities such as: Railway junction, small industries, educational institutes, health care centre, and offices/trust in Loharu, Bhiwani, Haryana, India. The anticipated daily unmet load demand is 8343kWh/day. The objective function is to minimize the expenses and emissions by integrating renewable energy sources, including a 100-kW wind turbine, a 55-kW Solar PV, a 500-kW bidirectional converter, and a Hydrogen tank of 100 kg into the existing grid. The HOMER Pro program is utilized to simulate the system. The large amount of expenses includes the net present cost, the levelized cost of energy, operating expenses, and capital expenditures, which are reduced annually. According to the environmental investigation, there has been a significant reduction in annual emissions included in this proposal, which are Carbon dioxide (CO_2), Carbon monoxide (CO), unburned hydrocarbons (UHC), particulate matter (PM), sulfur dioxide (SO_2), and Nitrogen oxides (NO_x) in kg/year, respectively. Overall, the findings illustrate that the proposed hybrid system is a realistic and environmentally friendly approach to increasing rural energy availability while lowering operating expenses and environmental emissions.

Keywords Economic · Eco-friendly · HOMER pro · Hybrid microgrid · Loharu

S. Singh (✉)
Department of Electrical Engineering, UIET, MDU, Rohtak, India
e-mail: surender.b1985@gmail.com

S. Kundu et al. (eds.), *Hybrid Renewable Power Infrastructure for Sustainable Electric Vehicle Development*, Studies in Infrastructure and Control,
https://doi.org/10.1007/978-981-95-1323-9_11

1 Introduction

The world's population is growing fast, but the supply of energy needed to maintain economic and social growth is still inadequate, suggesting that a significant percentage of the population lacks access to clean energy. The industrial revolution and centuries of consumption have been powered by coal and oil, which have long been the primary energy sources. Fossil fuels are restricted, though, and are bound to run out. Renewable energy has drawn attention as a potential substitute in the last decade. Microgrids are currently making progress from laboratory benches and pilot demonstration sites into commercial markets thanks to advancements in technology, declining costs, a track record of success, and increased awareness of their advantages. They are used to increase electrical grid resilience and dependability, control the addition of distributed clean energy resources, such as solar photovoltaic (PV) and wind power, to lower emissions from fossil fuels, and supply electricity in places without centralized electrical infrastructure [1]. Initially, a load flow study is required to meet the unmet load demand of a particular remote area. The load demand for estimation is categorized into balanced and unbalanced radial distribution systems that take weather conditions into account. The impact of ambient branch resistance circumstances is taken into account in order to pinpoint precise power losses and improve the effectiveness of unbalanced load flow solutions. This aims to solve unbalanced radial distribution networks while taking the weather into account. When impedances are seen as a fixed parameter rather than weather-sensitive, a discernible shift in power loss, resistance, and node voltages is seen when the findings are contrasted with traditional load flow analysis of unbalanced radial distribution systems [2]. Shifting the transition from non-sustainable fossil fuels to renewable energy sources provides numerous benefits and resolves several global challenges, including:

i. Renewable energy sources, including solar and wind, are essential for lowering greenhouse gas emissions, preventing global warming, and supporting environmental sustainability [3, 5]. Renewable energy sources have less of an impact on the environment than fossil fuels, which deteriorate the quality of the air and water, resulting in cleaner air and healthier ecosystems [6, 8].
ii. Renewable energy solutions offer stable, long-term energy costs, which over time lead to economic savings, regardless of the substantial initial investment [9, 11]. By reducing reliance on fossil fuels, renewable energy adoption helps to mitigate the risks associated with global instability and fluctuating fuel prices [12, 14].
iii. Access to energy is a problem in many places, especially in developing nations. Technologies for decentralized renewable energy raise the standard of living and encourage energy equity [15, 17]. The world is committing to achieving net-zero emissions by 2050, which includes major expenditures in renewable energy and a reduction in reliance on fossil fuels.

Under the Rajiv Gandhi Gramin Vidyuti Karan Yojna, the DHBVN has submitted several proposals worth approximately Rs. 100 crores for each district to the Indian government for approval in order to offer free connections to rural residents who are below the poverty line. The Indian government has already approved plans worth Rs. 29 crores for the districts of Sirsa and Bhiwani [18]. The research proposal focuses on designing and optimizing a hybrid energy system for the minimization of the objective function of Loharu location ($28^\circ 25.6^{1}$ N, $25^\circ 48.2^{1}$ E) in Loharu, Bhiwani District, Haryana, India, as shown in Figure 1.

A reliable plan for an independent Distributed Generation (DG) system that uses solar photovoltaic (PV), wind, and battery storage to support off-grid or remote applications. An adaptive controller that maximizes synchronization between Renewable Energy Sources (RESs) and the load is built using the Adaptive Normalized Least Mean Absolute Third method. This improves the system's overall reliability and Power Quality [19]. The investigation examines deeply at the challenges of incorporating renewable energy sources into distribution system architectures. It emphasizes the vitality of this integration for modernizing the energy system and achieving environmental objectives. The investigation examines compatibility challenges, overall performance indicators, and the particular challenges that on-grid power structures encounter. It also provides a comprehensive evaluation of the features of several renewable energy sources, hybrid systems, and technology from the solar, wind, battery, and biomass sources [20]. The research addresses the energy needs of the domestic and commercial sectors, providing a customized energy optimization strategy tailored to the village's unique requirements. The present research examines extensively at the challenges of incorporating renewable energy sources into distribution system architectures. It emphasizes the significance of this integration for modernizing the energy system and achieving environmental goals. The investigation examines compatibility difficulties, overall performance indicators, and the particular challenges that on-grid power structures face. In order to highlight the importance of creative solutions within the scattered technology environment, it also provides a comprehensive evaluation of the characteristics of different hybrid systems of renewable energy sources, including technology from the solar, wind, battery, and biomass sectors [21]. The study proposes a distributed generation (DG) system's synchronization and control approach. Additionally, a complex coefficient filter-based enhanced phase-locked loop (ICCF-PLL) control method is explained. Under recurrent conditions of renewable energy sources (RESs) and load disturbance, an ICCF-PLL control technique minimizes oscillations, minimizes steady-state error, and offers a quick dynamic response [22].

An autonomous microgrid (MG) is being developed utilizing a wind-driven self-excited induction generator (SEIG), solar photovoltaic (SPV), and energy storage system (ESS). To identify phase angle and provide unit templates for a power quality management scheme in an autonomous MG, an improved phase-locked loop (VGC-IPLL) based on a variable gain controller is employed [23-25].

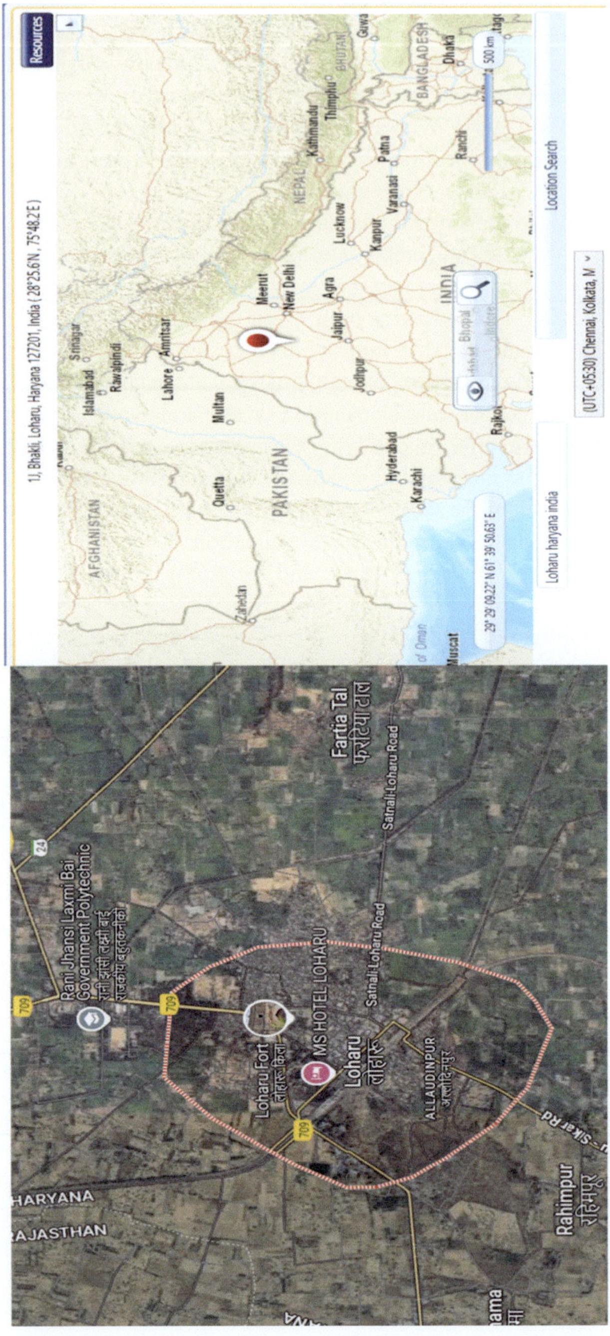

Fig. 1 Location of the hybrid system at Loharu

The proposed investigation highlights the considerable reductions in economic and environmental emissions achieved by the transition from non-sustainable fossil fuels to renewable energy sources:

i. A healthier environment is eventually promoted by the significant reduction of emissions caused by the integration of renewable energy sources, which also helps to mitigate global warming and improve air quality.
ii. Although the hybrid system involves high initial investment costs, it proves to be more cost-effective in the long run by significantly reducing overall expenses. Additionally, placing renewable sources close to the load improves the sustainability and reliability of the power supply.
iii. In addition to cutting costs and reducing emissions, the proposed hybrid energy system is designed to meet the substantial daily energy demand of 8341.69 kWh for Loharu, Bhiwani District, Haryana, India, thereby ensuring an uninterrupted and sustainable power supply.

The schematic diagram for the proposed hybrid system is demonstrated in Fig. 2 by integrating renewable sources into the conventional energy source for essential infrastructure in Loharu, Bhiwani District, Haryana.

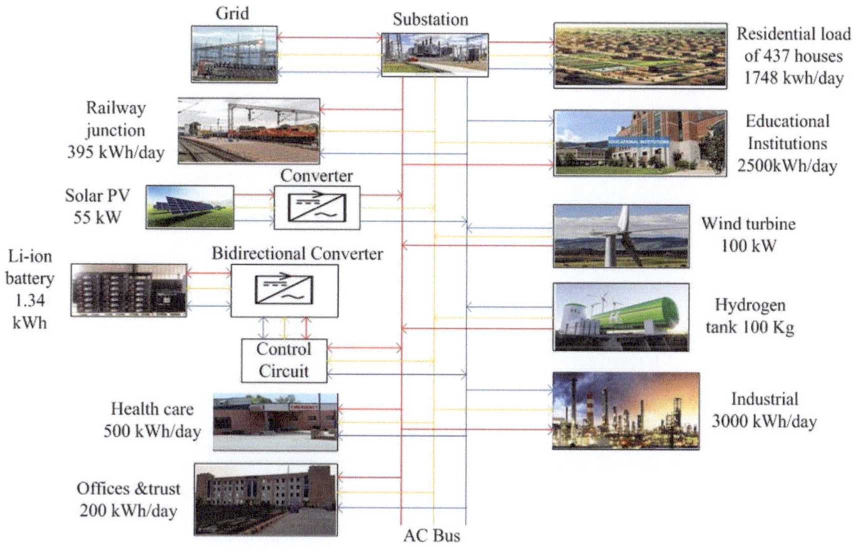

Fig. 2 Schematic diagram for the proposed hybrid system

2 Methodology

The objective function for the proposed hybrid system is a combination of expenses and emissions as follows:

$$\text{Obj.}(F) = \min.\left[C_{\text{expenses}} + C_{\text{emission}}\right] \tag{1}$$

$$C_{\text{expenses}} = \text{NPC} + L_{\text{COE}} + \text{OC} \tag{2}$$

where NPC represents the net present cost, OC represents the operating cost, and LCOE represents the levelized cost of energy.

NPC: A component's net present cost, also known as life-cycle cost, is the present value of all installation and running expenses throughout the project, less the present value of any income made during that period. The NPC, CRF, and L_{COE} [26, 27] are estimated by HOMER for both the system as a whole and for each component as follows:

$$\text{NPC} = \frac{\text{TAC}}{\text{CRF}(i, Rprj)} \tag{3}$$

$$\text{CRF(i, n)} = \frac{\text{i} \times (1 + \text{i})^{n}}{(1 + \text{i})^{\text{n}} - 1} \tag{4}$$

$$\text{L}_{\text{COE}} = \text{TAC/Total annual energy supplied} \tag{5}$$

where CRF represents the capital recovery factor, TAC is the total annualized cost, (i) is the real discount rate, and n is the number of years.

Emission Equation: The emission of the pollutant due to power generation from fossil fuel [27].

$$E(emission) = \sum\nolimits_{i=1}^{n} \text{a(i).P(generation), i(t)} \tag{6}$$

where $a(i)$ represents the emission factor of the ith generation source, P (generation) represents the power generation, and $i(t)$ represents the power output at the ith generation.

Flowchart: The flowchart for the proposed hybrid system to achieve the minimization of the objective function of expenses and emissions is elaborated in Fig. 3.

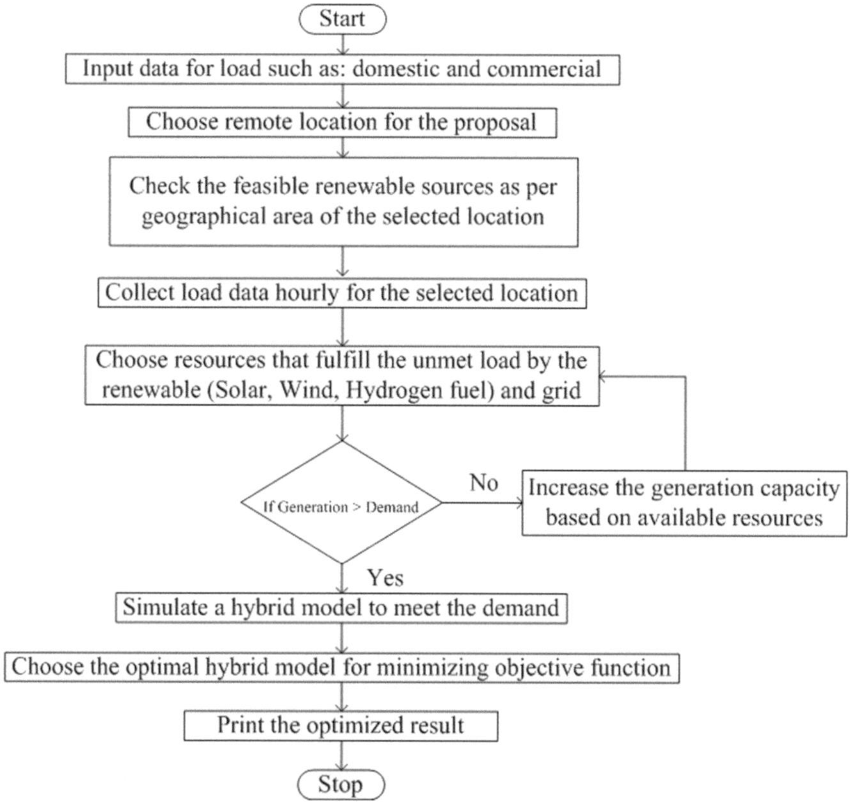

Fig. 3 Flow chart of the proposed hybrid system

3 Results and Discussion

The energy requirement for Loharu, Bhiwani District, Haryana, India, has been assessed through primary survey data, which includes information on various load categories, such as domestic and commercial, for the rural enterprises. The assessment helps in understanding the energy consumption patterns and load profiles for the electrification of the village of 437 houses. The daily domestic load requirement is 1748 kWh/day. The commercial load, including Railway junction (395 kWh/day), Health Care (500 kWh/day), Educational Institute (2500 kWh/day), Industrial (3000 kWh/day), Offices and trust (200 kWh/day), is 6595 kWh/day. Therefore, the total load demand is 8343 kWh/day. The domestic load includes typical household electricity consumption such as lighting, fans, televisions, mobile charging points, and water pumps for drinking water supply, and the rated capacity of appliances [28] shown in Table 1.

Table 1 Daily load requirement

Domestic appliances	Rated capacity of the appliance (W)	Quantity of domestic appliances (N)	Usage/day in hours	Consumption/day (kWh) = (Load × Usage per Day)/ 1000
Light (CFL)	15	10	6	0.9
Ceiling Fan	30	5	8	1.2
TV	80	1	2	0.16
Table fan	20	1	2	0.04
Power socket	500	3	1	1.5
Others	308	1	1	0.308
Total	953	21	20	400.26

Designing the model to meet the unmet demand at the remote location can be achieved in two ways: Case 1 and Case 2. The schematic diagram for both Cases is shown in Fig. 4.

Case 1: In this case, include meeting the load demand by Solar PV and Grid both, and simulate the system with the help of HOMER by achieving the minimization of the objective function, which is a combination of expenses and emissions. The Solar PV, Grid, load demand, converter, and simulation results are elaborated as follows:

Solar PV: A panel of a flat plate, rated Capacity 680.08 kW, temperature Coefficient -0.4100, operating temperature 45 °C, and efficiency of 17.30% is utilized in Case 1.

Converter: Bi-directional inverters 500 KW in the 100 series are especially made for energy storage applications in microgrids and grid-tied systems. Its inverter has two inputs. It can be configured as a battery or PV with a single input configuration of 280–600 Vdc/320A dc with maximum power point tracking. The converter is designed to interface between parallel AC and DC bus layouts. Its dual roles as

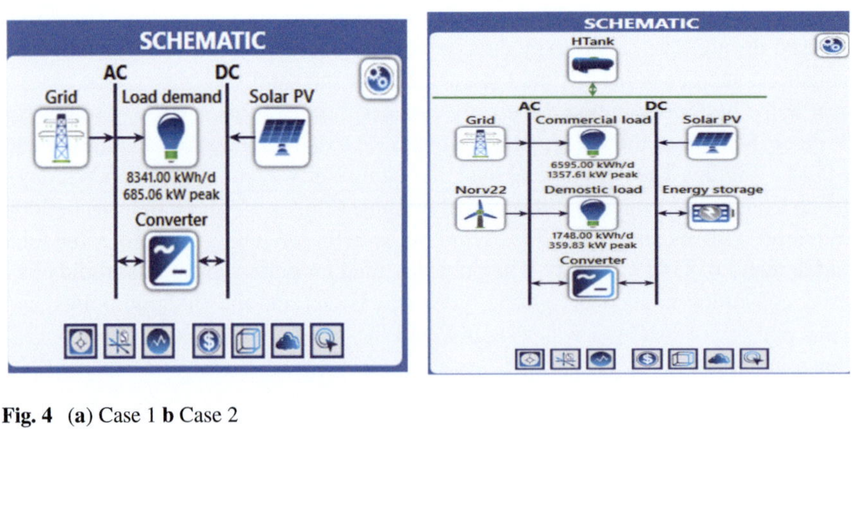

Fig. 4 (**a**) Case 1 **b** Case 2

an AC-DC converter and DC-AC inverter allow for effective bidirectional power conversion and guarantee the smooth integration of energy storage or renewable energy sources with the grid. The converter maintains steady voltage and frequency while optimizing power flow. It ensures efficient energy conversion, optimizing the overall performance of the solar energy system. The converter equation is given as under:

$$P_{\mathrm{DCtoAC}} = \gamma_{\mathrm{converter}} \times P_{DC} \tag{7}$$

where P_{DCtoAC} represents the converter DC to AC power at time t, $\gamma_{converter}$ is the efficiency of the converter and P_{DC} represents the DC power output of the converter.

Grid: The Grid meets the load demand monthly as well as annually, as shown in Table 2. Here, it is observed that a part of the energy demand is shared by the Solar PV, but the energy sold back to the Grid is zero. As a result of these few expenses, huge emissions were minimized in this case.

The estimated pollutant emission in this case is presented in Table 3.

The simulation results in this case are presented in Fig. 5. The power generated by Solar PV is 30613 kWh/year, which is much less than the total energy demands. Due to the small penetration of renewable sources into the existing Grid, it saves energy expenses and reduces emissions of pollutants.

The variation of primary load, Solar power penetration, rectifier power output, and inverter power output is demonstrated for a weak system in Fig. 6. The penetration of Solar PV is limited by geographical area and solar radiation during daytime, including cloud shading.

Table 2 Load demand monthly as well as annually for the case 1

Month	Energy purchased (kWh)	Energy sold (kWh)	Net energy purchased (kWh)	Peak load (kW)
January	2,53,618	0	2,53,618	668
February	2,24,366	0	2,24,366	663
March	2,62,755	0	2,62,755	664
April	2,49,356	0	2,49,356	670
May	2,50,925	0	2,50,925	624
June	2,50,958	0	2,50,958	671
July	2,53,273	0	2,53,273	678
August	2,64,947	0	2,64,947	623
September	2,50,609	0	2,50,609	609
October	2,53,607	0	2,53,607	581
November	2,43,754	0	2,43,754	641
December	2,56,755	0	2,56,755	635
Annual	30,14,923	0	30,14,923	678

Table 3 Pollutant emissions in case 1

Emission pollutant	Quantity (Kg/year)
Carbon dioxide	19,05,432
Carbon monoxide	0
Unburned hydrocarbons	0
Particulate matter	0
Sulfur dioxide	8261
Nitrogen oxides	4040

Case 2: Here, the objective function—a combination of expenses and emissions—must be minimized in order to simulate the system using HOMER and include achieving the load demand from both the grid and integration of renewable sources as Solar PV, Wind turbine, Storage, and a bidirectional converter to meet the domestic and commercial load demand. The results of the simulation, load demand, Solar PV, Wind turbine, Storage, and a bidirectional converter are detailed as follows:

Grid: As demonstrated in Table 4, the wind turbine fulfills the load demand for every month and annually. Here, it can be observed that while the solar PV shares a portion of the energy demand, a huge amount of energy is sold back to the grid. As the small amount of energy generated by the Solar PV is due to the geographical area, the wind turbine shares approximately 96% of the total demand due to this remote location. Henceforth, a huge saving in expense and significant emissions were reduced in this case.

Wind turbine: A wind turbine of 100 kW rated capacity is utilized for the proposed project. It is made with cutting-edge technology to maximize performance and generate energy efficiently in large-scale wind power applications. Reliable and sustainable energy generation is offered by this turbine, which may be installed both onshore and offshore. The wind energy generation equation is given as under:

$$\text{Ewind(t)} = \frac{1}{2}.\rho.\text{A(rotor)}.\text{V}^3.\text{k(wind)} \tag{8}$$

where Ewind(t) is the energy generated by the wind turbine at time t, ρ represents air density, A(rotor) is the area swept by the rotor, V represents the wind speed at time t and k(wind) represents the efficiency of the wind turbine. Monthly average wind speed data at 50 m height, including wind speed profile variations, derived from NASA's POWER database, as shown in Fig. 7.

The specification for the wind turbine is given in Table 5.

Battery Storage: In case of energy storage having a nominal capacity of 1.34 kWh, the lifetime becomes 36,000 kWh, bus voltage 8.40 V, and the expected life becomes 23 years. The minimum state of discharge considered here is 10% of the rated capacity, and the maximum state of charge considered is 90% of the rated capacity. The charging equation follows as under:

$$E_{\text{Charge}} = \gamma_{\text{Charge}} \times P_{\text{Charge}} \tag{9}$$

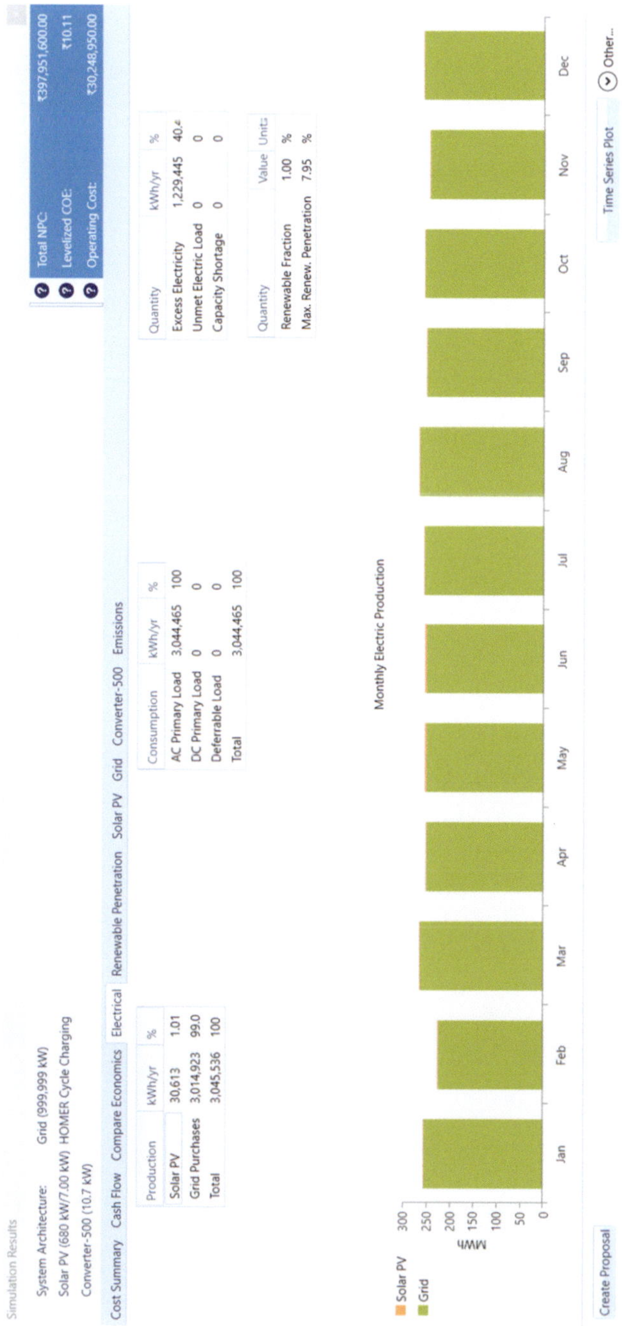

Fig. 5 Simulation results in case 1

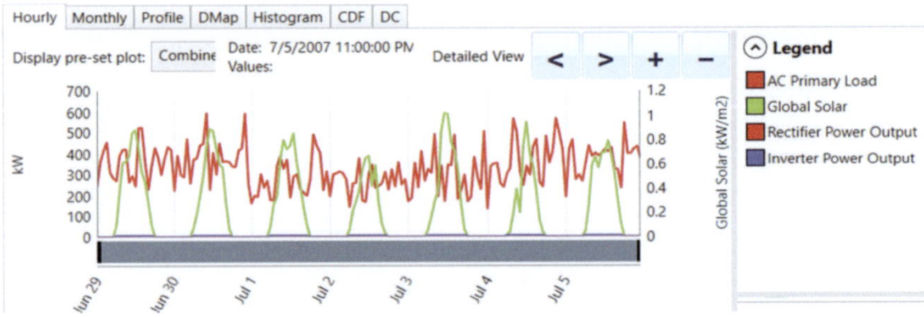

Fig. 6 Variation of primary load, Solar power penetration, rectifier power output, and inverter power output

Table 4 Load demand monthly as well as annually for the case 2

Month	Energy purchased (kWh)	Energy sold (kWh)	Net energy purchased (kWh)	Peak load (kW)
January	94,804	10,43,446	−9,48,642	1544
February	80,260	10,65,214	−9,84,954	1463
March	81,075	12,42,041	−11,60,966	1551
April	58,252	14,95,793	−14,37,541	1198
May	40,870	22,15,874	−21,75,004	943
June	35,325	23,27,006	−22,91,681	1002
July	34,455	21,46,347	−21,11,893	819
August	42,590	19,54,291	−19,11,702	1042
September	50,081	18,28,913	−17,78,831	1002
October	48,142	18,84,021	−18,35,879	1326
November	47,918	19,20,421	−18,72,503	1257
December	80,456	15,28,951	−14,48,495	1380
Annual	6,94,228	2,06,52,318	−1,99,58,090	1551

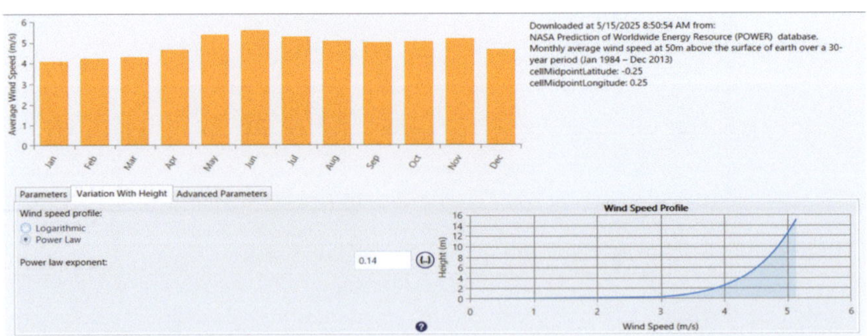

Fig. 7 Monthly average wind speed data

Table 5 Specification for the wind turbine

Wind specification	Performance
Maximum output	10,000 kW
Hours of operation	6887 h/yr
Levelized cost	0.00121 ₹/kWh
Mean output	2615 kW
Capacity factor	26.10%
Total production	22,905,073 kWh/yr

The discharging equation is calculated as:

$$E_{\text{Discharge}} = \left\{ \frac{1}{\gamma_{\text{Discharge}}} \right\} \times P_{\text{Discharge}} \tag{10}$$

The State of Charge (SOC) equation is given as:

$$\text{SOC (t)} = \text{SOC (t} - 1) \frac{E(\text{charge})t - E(\text{discharge})t}{C(bat)} \tag{11}$$

where E_{Charge} represents the amount of energy stored in the battery, γ_{Charge} is the efficiency of the battery and P_{Charge} is the power capacity required to be stored, $E_{\text{Discharge}}$ represents the energy discharge of the battery, $P_{\text{Discharge}}$ represents the power discharge of the battery, and SOC (t) is the state of charge at time t in the battery.

Solar PV: Generic solar modules are used to generate electricity in the 55-kW photovoltaic system under examination. Here, geographical limitations make solar energy a viable solution to meet demand. As a result, flat solar PV is shown in the simulation model. The charge controller, which is integrated into the system, is essential for maximizing energy flow between the solar array and the load or storage systems. Monthly average solar global horizontal irradiance (GHI) data, including daily radiation and clearness index, derived from NASA's POWER database, as in Fig. 8.

Converter: The specification for the converter is given in Table 6.

Hydrogen tank: The specification for the Hydrogen tank is given in Table 7.

Emission: The emission pollutants in Case 2 are depicted in Table 8.

Figure 9 displays the simulation findings, and it is observed that 101,792 kWh of power is produced by solar photovoltaics annually, which is significantly less than the entire energy requirement. Because of the unlimited integration of wind turbines into the current grid, energy expenses are decreased, and pollution emissions are decreased.

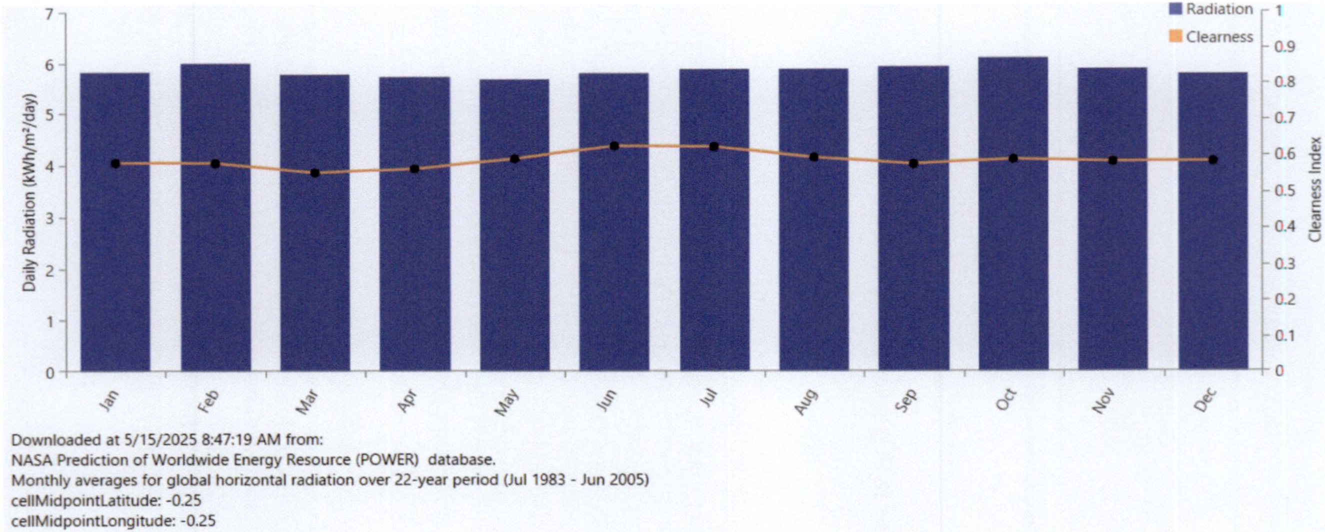

Fig. 8 Monthly average solar global horizontal irradiance (GHI) data

Table 6 Specification for the converter

Converter specification	Inverter	Rectifier
Capacity	50 kW	50 kW
Mean output	11.2 kW	0
Capacity factor	22.40%	0
Hours of operation	4380 h/yr	0
Energy out	98,213 kWh/yr	0
energy in	101,775 kWh/yr	0
Losses	3562 kWh/yr	0

Table 7 Specification for the Hydrogen tank

Hydrogen tank	Quantity
Hydrogen storage capacity	100 kg
Energy storage capacity	3333 kWh
Tank autonomy	9.59 h

Table 8 Emission pollutants in Case 2

Emission pollutant	Quantity (Kg/year)
Carbon dioxide	4,38,750
Carbon monoxide	0
Unburned hydrocarbons	0
Particulate matter	0
Sulfur dioxide	1,902
Nitrogen oxides	930

For a weak system, Fig. 10 shows how the primary load, load served, wind turbine output, stored hydrogen in the tank, and inverter power output vary. Geographical location and daytime solar radiation, including cloud shading, are the main factors limiting the adoption of solar photovoltaics. But the wind turbine generates the maximum power to meet the load demand during the day and night in this remote location.

Fig. 9 Simulation results for Case 2

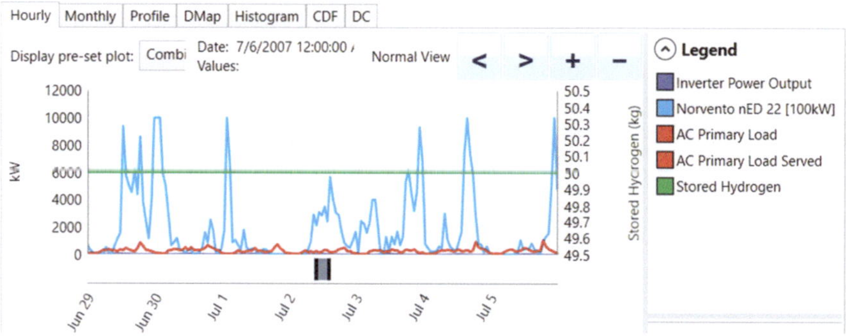

Fig. 10 Variation of renewable source output and primary load served

4 Conclusion

The Case 2 energy system demonstrates significant economic and environmental benefits over the Case 1 system. From an economic perspective, the NPC saving of ₹ 1,805,259,000 annually, L_{COE} saving 5.89 ₹/kWh, and OC saving ₹139,683,800 in Case2 energy expenses annually. However, Case 2 has a higher initial investment cost, which is compensated for by the short duration by saving in energy expenses. The environmental impact in Case 2 of the system leads to substantial reductions in pollutant emissions as: CO_2 reduces 1,466,682 kg/year, SO_2 reduces 6359 kg/year, and NO_x reduces 3110 kg/year, respectively. The reductions in expenses and enhancing an environmental benefit while implementing Case 2, which contributes to significant savings in cost and cleaner air by reducing emission pollutants.

Acknowledgements Declaration of Competing Interest The authors declare that this chapter did not receive any specific grant from funding agencies in the public, commercial, or not-for-profit sectors.

Data Availability No data were used for the research described in this chapter.

References

1. Hirsch A, Parag Y, GuerreroJ (2018) Microgrids: a review of technologies, key drivers, and outstanding issues. Renew Sustain Energy Rev 90:402–411. https://doi.org/10.1016/j.rser.2018.03.040
2. Singh S, Singh VR, Ranjan R, Swapnil S (2019) Advanced load flow analysis for unbalanced radial distribution system incorporating weather conditions. J Inst Eng (India): Ser B 100:551–560
3. International Energy Agency (IEA) (2022) World Energy Outlook 2022. International Energy Agency, Paris, France. Available: https://www.iea.org/reports/world-energy-outlook-2022
4. REN21 Secretariat (2022) Renewables 2022 Global Status Report. REN21 Secretariat, Paris, France. ISBN: 978-3-948393-04-5. Available: https://www.ren21.net/gsr-2022/

5. Intergovernmental Panel on Climate Change (IPCC) (2021) Climate Change 2021: the physical science basis. Cambridge University Press, Cambridge, U.K. ISBN: 978-1-00-915789-6. https://doi.org/10.1017/9781009157896. Available: https://www.ipcc.ch/report/ar6/wg1/
6. World Health Organization (WHO) (2018) Air pollution and child health: prescribing clean air. World Health Organization, Geneva, Switzerland. ISBN: 978-92-4-151534-0. Available: https://www.who.int/publications/i/item/air-pollution-and-child-health
7. United Nations Environment Programme (UNEP) (2019) Global environment outlook 6. UNEP, Nairobi, Kenya. ISBN: 978-1-108-62714-6. https://doi.org/10.1017/9781108627146. Available: https://www.unep.org/resources/global-environment-outlook-6
8. Jacobson MZ et al (2020) A roadmap for 100% clean, renewable energy worldwide. Energy Environ Sci 13(3):793–802. https://doi.org/10.1039/C9EE01208A. Available: https://pubs.rsc.org/en/content/articlelanding/2020/ee/c9ee01208a
9. BP (2023) Statistical Review of World Energy 2023. BP, London, U.K. Available: https://www.bp.com/en/global/corporate/energy-economics/statistical-review-of-world-energy.html
10. Ministry of New and Renewable Energy (MNRE) (2024) Government of India, Annual Report 2023–24. MNRE. New Delhi, India. Available: https://mnre.gov.in/
11. International Renewable Energy Agency (IRENA) (2022) Renewable energy market analysis: India and Southeast Asia. IRENA, Abu Dhabi, UAE. Available: https://www.irena.org/publications
12. Goldemberg J (2008) The case for renewable energies. Energy Policy 36(11):3683–3687. https://doi.org/10.1016/j.enpol.2008.07.008. Available: https://www.sciencedirect.com/science/article/abs/pii/S0301421508003679
13. United Nations Development Programme (UNDP) (2022) Sustainable Development Goals Report 2022. UNDP, New York, NY, USA. Available: https://unstats.un.org/sdgs/report/2022
14. NITI Aayog (2023) India Energy Security Scenarios 2047. Government of India, New Delhi, India. Available: https://www.niti.gov.in
15. World Bank (2023) Access to electricity: progress and policies. World Bank Group, Washington, DC, USA. Available: https://www.worldbank.org/en/topic/energy
16. Blakers T, Lu B, Stocks M (2019) 100% renewable electricity in Australia. Energy 133:471–482. https://doi.org/10.1016/j.energy.2017.05.168. Available: https://www.sciencedirect.com/science/article/pii/S036054421731088X
17. Government of India (2023) National Electricity Plan 2023, Central Electricity Authority, New Delhi, India. Available: https://cea.nic.in/
18. Dakshin Haryana Bijli Vitran Nigam (2025) Information Handbook 14 May 2025. https://dhbvn.org.in/staticContent/information/handbook.pdf
19. Kundu S, Giri AK, Kadiyan S, Singh S (2025) Power management and adaptive control method for a standalone distributed generation system under fault, DC-offset, and dynamic environments. Int J Ambient Energy 46(1). https://doi.org/10.1080/01430750.2025.2452245
20. Singh S (2025) Renewable-energy-based EV charging infrastructures. In: Giri AK, Singh M (eds) Electric vehicle charging infrastructures and its challenges, studies in infrastructure and control. Springer, Singapore, pp 71–86. https://doi.org/10.1007/978-981-96-0361-9_4
21. Singh S, Singh S (2024) Advancements and challenges in integrating renewable energy sources into distribution grid systems: a comprehensive review. J Energy Resour Technol 146(9). https://doi.org/10.1115/1.4065503
22. Kundu S, Singh M, Giri AK (2024) Synchronization and control of WECS-SPV-BSS-based distributed generation system using ICCF-PLL control approach. Electr Power Syst Res 226
23. Kundu S, Singh M, Giri AK (2023) Implementation of variable gain controller based improved phase locked loop approach to enhance power quality in autonomous microgrid. Int J Numer Model: Electron Netw, Devices Fields 36(5)
24. Kundu S, Singh M, Giri AK et al (2024) Robust and fast control approach for islanded microgrid system and EV charging station applications. Electr Eng. https://doi.org/10.1007/s00202-024-02291-1
25. Kundu S, Singh M, Giri AK (2024) SPV-wind-BES-based islanded electrical supply system for remote applications with power quality enhancement. Electr Eng 106:279–294. https://doi.org/10.1007/s00202-023-01979-0

26. Rezzouk H, Mellit A (2015) Feasibility study and sensitivity analysis of a stand-alone photovoltaic-diesel-battery hybrid energy system in the north of Algeria. Renew Sustain Energy Rev 43:1134–1150. https://doi.org/10.1016/j.rser.2014.11.103. Available: https://www.sciencedirect.com/science/article/abs/pii/S1364032114009646
27. HOMER Energy. Calculation of Emissions using HOMER Pro Software. Available: https://homerenergy.com/products/pro/docs/3.15/how_homer_calculates_emissions.html
28. Bureau of Energy Efficiency (BEE) (2023) Government of India. Standards & Labeling Programme. Available: https://beeindia.gov.in/en/programmes/standards-labeling. Accessed 29 Mar 2025

Chapter 12
Future Trends and Innovation in EV Charging

Deepesh Sharma, Mohit Kumar, and Atma Ram

Abstract The decarbonization of the transportation sector necessitates the accelerated deployment of electric vehicles (EVs) and corresponding charging infrastructure. This study presents a comprehensive analysis of contemporary EV charging technologies, emphasizing the transition from conventional Level 1–3 systems to advanced modalities such as ultra-fast charging (>350 kW), inductive and dynamic wireless power transfer (DWPT), and vehicle-to-grid (V2G) bi-directional energy systems. The integration of renewable energy sources with charging stations, particularly solar and wind, facilitated through AI/ML-driven demand forecasting, dynamic pricing models, and IoT-enabled real-time grid communication, is also examined. The research further explores smart grid architectures, including dynamic line rating systems, load balancing protocols, and ancillary service support, which are critical to ensuring power system stability. Control strategies such as V1G unidirectional control, master–slave coordination, internal model control (IMC), and decentralized energy resource (DER) optimization are analyzed for their role in improving energy efficiency and reducing charging latency. The Chapter contributes to the evolving discourse on sustainable electromobility by highlighting the synergistic potential of cyber-physical integration, intelligent infrastructure planning, and user-centric EV charging design. These insights are intended to guide policymakers, utility providers, and manufacturers in developing resilient and scalable charging ecosystems.

D. Sharma (✉) · M. Kumar
Electrical Engineering Department, Deenbandhu Chhotu Ram University of Science and Technology, Murthal, Sonepat, Haryana, India
e-mail: deepeshsharma.ee@dcrustm.org

M. Kumar
e-mail: 20001902006mohitkumar@dcrustm.org

A. Ram
J C Bose University of Science and Technology, YMCA, Faridabad, India
e-mail: atmaram@jcboseust.ac.in

S. Kundu et al. (eds.), *Hybrid Renewable Power Infrastructure for Sustainable Electric Vehicle Development*, Studies in Infrastructure and Control,
https://doi.org/10.1007/978-981-95-1323-9_12

1 Introduction

Transportation is changing due to the desire worldwide to limit greenhouse gas emissions, depend less on fossil fuels, and fight urban air pollution. The main driving factor behind the change is the fast uptake of electric vehicles (EVs), which are cleaner and more fuel-efficient than combustion engine vehicles. But the EV market's expansion is largely controlled by how widely and how well supporting structures, especially EV charging systems, are available.

At present, charging EVs can be done with Level 1 (slow), Level 2 (AC fast), or Level 3 (DC fast) technology, and their outputs and uses depend on each type. Even though they made EV ownership possible in the early days, these technologies cause delays due to slow charging, deter the grid, are not equally available, and may not fit all EV and charger types [1, 2]. Because of these limitations, researchers and developers are exploring new and larger ways to charge electric vehicles.

To solve these problems, new trends in EV charging mainly use technologies like ultra-fast charging, wireless (inductive) solutions, vehicle-to-grid (V2G) integration, battery swapping, and support from AI and IoT [3–6]. In order to reduce charging time, ultra-fast charging brings power above 350 kW [7]. Many people are noticing wireless charging for its ease of providing electricity to cars without the use of charging cables [8]. In addition, V2G technology means that EVs can take energy from and give energy to the grid, which strengthens the grid's flexibility and reduces risks of failure [9].

Thanks to EV infrastructure and smart grids coming together, there is now a better ability to manage power loads, check usage in real time, and adjust demand when needed, key aspects in making sure losses and grid issues are reduced with more EVs. As well as being technical, these developments also rely on new policies, as governments and regulators create laws, incentives, and funds to speed up deploying EV charging points everywhere.

Due to these advancements, the purpose of this paper is to analyze the EV charging market as it is now, highlight the most promising updates, and look ahead at what might change the electric car charging experience for people and energy companies. Investigating these aspects, the study helps move forward debates in sustainable transportation and offers insights on what is needed to plan an effective EV infrastructure.

2 Future Trends and Innovations in EV Charging

The field of electric vehicle charging stands at the threshold of transformative change since technology advances and rising customer needs and sustainability requirements demand thorough research into new innovations and developing patterns [10].

Time limitations and charging anxiety must be resolved for EV adoption to gain mass popularity since swift recharging systems represent a critical development [11, 12]. A sustainable transportation system based on electrified solutions shows great promise in reducing climate change effects, which requires building successful, affordable, rapid charging networks for EV commercial success [12]. The EV industry transition requires massive infrastructure investments because growing charging demands offset numerous benefits of EV adoption, including environmental and economic advantages [13]. Building efficient and dependable fast charging stations near each other will provide EVs with unrestricted driving distances [12]. High-voltage charging techniques require advanced thermal management systems to make fast charging technologies work into the future [14].

When implementing EV charging infrastructure into smart grids, several challenges emerge together with beneficial potential, including infrastructure planning and power quality, network loading and ancillary services, and reactive power support and load balance requirements [12]. The management of high EV charging volumes through smart charging systems becomes crucial because it decreases electrical grid disturbances, as documented in [11, 15]. The system called Vehicle-to-grid (V2G) enables electric vehicles to interact as suppliers by exchanging stored energy with the electricity grid [12]. The European Commission targets a major percentage of total energy consumption from renewable sources while highlighting the necessity to unite EV charging capabilities with renewable energy infrastructure [16]. Distribution-based power generation systems produce challenges regarding power distribution patterns, together with operational voltage standards and fault current limitations [17]. The smart grid integration of electric vehicles requires proper charging management techniques that create multiple opportunities, primarily through vehicle-to-grid technology and renewable energy stability solutions [18]. The combination of innovative products with intelligent monitoring and control services, communication protocols, and self-healing technologies makes up smart grids as they get integrated into utility processes and systems [19]. The influence of dynamic line rating implementation spreads to affect electricity market price areas, according to research found in [19, 20].

Dynamic Line Rating systems develop as essential elements of smart grid technologies through improved monitoring and communication technologies that enable implementation and exploitation [19]. All power system actors, including network operators, along with market players and producers or consumers, will enhance dynamic line rating by enabling automatic control through available information about variable constraints [19]. The implementation of dynamic line rating systems includes a cost-effective method to enhance grid capacity through real-time monitoring, which allows transmission limit adjustments [16]. Static line ratings establish improperly low transfer threshold values, but dynamic line rating delivers multiple network benefits, which include improved transmission capacity management, cost savings, and reduced network infrastructure requirements [21]. The dense population and intensive land utilization limit electrical line allocation areas, so dynamic rating systems help prevent new line constructions [16]. The application assessment for DLR needs to be performed individually because both weather variations and transmission limitations affect the technology [19].

Inductive charging and resonant inductive charging represent two wireless charging technologies that are gaining popularity as easy-to-use charging solutions instead of traditional plug-in charging processes [16]. The market is developing automated charging solutions with robotic charging arms and autonomous charging vehicles to boost the convenience of EV charging [22]. The improvements become essential for autonomous car technology and shared mobility systems because users need effortless charging systems. Sustainable transportation targets demand an absolute unification between renewable energy systems with EV charging facilities, so solar-powered charging stations and wind-powered charging stations emerge increasingly often [23]. The integration between renewable energy systems and EV charging technology allows both transportation decarbonization and energy independence to succeed simultaneously.

Dynamic charging systems function as a solution that addresses both charging station and conductive charging solution constraints [24]. Dynamic charging systems enable automobiles with electric propulsion to recharge even when they are moving thus enabling extended driving ranges [25, 26]. The dynamic wireless charging solution provides an accessible answer to resolve the problems found in EV charging systems by extending vehicle range and shortening charging periods [27, 28]. EV cruising distance increases while the size of energy storage systems decreases using wireless charging technologies [29]. Electric vehicles benefit from dynamic wireless charging through their ability to both minimize and erase range anxiety effects, together with battery capacity reduction needs [30]. As we can see in Fig. 1: Future of EV Charging, which is helping us to understand methods that might be used in the future for EV charging.

1. Ultra-Fast Charging (UFC)
    - 10-min charging to 80% capacity.
    - EV charging technology depends on top-level thermal management in combination with battery management systems.
2. Wireless & Dynamic Charging
    - Wireless pads or coils have been installed within roads to enable vehicle charging.
    - Enables charging while driving (DWPT).
3. Vehicle-to-Grid (V2G) Integration
    - EVs act as mobile energy storage.
    - EV charging supports both grid stability functions and load balancing activities.

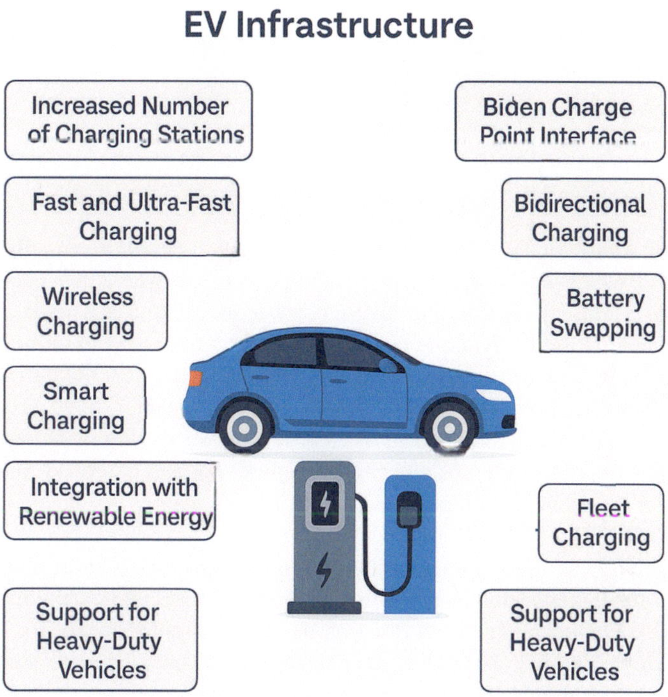

Fig. 1 Future of EV charging

4. AI-Driven Smart Charging
   - The system uses AI/ML to forecast when customers should charge their vehicles for maximum efficiency.
   - Minimizes costs and grid stress.
5. Standardized 'Plug & Charge'
   - Universal authentication/payment protocol.
   - Users should be able to begin charging without requiring any application interfaces or payment cards because the system works when simply plugged in.
6. Renewable Energy Integration
   - Charging synced with solar/wind availability.
   - Promotes sustainable energy use.

7. 5G & IoT-Enabled Infrastructure
   - Real-time data exchange between EVs and grid.
   - Improves charging coordination and efficiency.
8. Solid-State Batteries
   - Higher energy density, faster charging, safer.
   - Long-term game-changer in EV performance.

Dynamic charging systems enable longer battery operation combined with a smaller size, which leads to decreased maintenance expenses [24]. Presently, the installation expenses for dynamic charging systems remain at a high level [24]. Wide implementation of dynamic charging systems would lead to unlimited range for driving and eliminate electrical dangers [27]. Innovative fast inductive charging systems developed for static and on-route applications help solve problems by allowing wireless battery recharging of electric vehicles during brief periods [31]. Complexities related to EV costs during both purchase and extended maintenance periods deter extensive EV adoption, especially among nations that depend primarily on fossil fuels [32].

The development of sophisticated charging stations with fast charging capabilities is essential for EV adoption as it eliminates charging fears and expedites refueling periods [12]. Off-board chargers function as essential components for extending the reach of PEVs [33] [34]. The convenience of EV ownership depends on DC fast charging technology because it offers shorter charging times [35]. Extraneous charging facilities help EV owners gain more convenience and shorten waiting times during recharging sessions [36]. Power automation in electromobility gained momentum because EV adoption grows while the world needs better infrastructure systems [37]. Smart EV charging systems bring benefits to users along with advantages to operators and improvements to societal aspects [38]. Level 4 charging technology implements revolutionary EV infrastructure attributes that provide quick charging while implementing smart technological controls and adjustable power networks for various EV types. [39] The application of an internal model control system with synchronized decoupled controllers has allowed the development of a new EV fast charger, which supports AC and DC fast charging while facing hurdles regarding harmonic generation and control complexity and cost [40]. Instances of Level 2 charging require a near-208-to-240 V voltage input while being costlier than Level 1 charging methods [41]. Most current charging stations function as level-2 devices, while level-3 operators, known as DC fast chargers, offer fast charging alternatives. The main challenge for buying an EV arises from lengthy charging times associated with level-2 charging equipment [42]. More level-3 charging stations established in the market will drive faster EV adoption rates [43].

The development of reliable electric vehicle products drives fast growth in vehicle technology. The development of EV technologies receives support because it helps reduce carbon dioxide emissions and fights against fossil fuel pollution [44]. The charging station design includes features that allow for perfect integration with electrical grid networks while ensuring compatibility with upcoming technological breakthroughs. Rising charging technologies supply battery swapping alongside dynamic charging and inductive charging to supply quicker charging and improved flexibility and convenience [45]. Public support for the growth of the sector and EV owners comes from government programs, which offer tax advantages and cash-back benefits. The incentives provided by governments help decrease the cost of purchasing EVs, thus allowing more prospective customers to afford them. Advanced charging technology developments feature higher charging speeds together with smart charging features, as well as adopted charging standards, which enhance overall satisfaction throughout the user experience. EV charging solutions will evolve by providing functions that enhance owner convenience and by improving speed and grid connectivity to address existing limitations within the EV market [36]. Several new EV designs from manufacturers aim to fully support the EV market adoption [46].

Fast transportation electrification demands new charging infrastructure solutions because electric vehicle demand keeps escalating [47]. EV technology proliferation requires immediate action for convenient charging solutions because it will drive improvements in charging technologies while building new infrastructure. Toward successful transportation electrification, we need better battery technology and increased charging facility establishment [48]. The implementation of EV charging stations stands as a necessity for achieving transportation electrification because proper planning and financial investment create broad charging infrastructure availability [49]. Many countries have sped up the development of electric vehicles because it helps them decrease their reliance on oil and minimize environmental pollution [50]. Later developments in EV charging infrastructure will be defined by three technological trends, which include the combination of EV automobiles and smart power grids alongside renewable energy incorporation into charging stations and protocol standardization [51]. Electric cars are gaining more popularity because they present a useful solution to cut pollution and lower fuel dependence [52]. EV charging technology keeps advancing because researchers focus on three major improvements: faster charging rates and lower installation expenses, and enhanced comfort for users. EV market adoption will be extensive in the upcoming years, so EV manufacturers must make substantial investments to support the rise of EV ownership numbers [53, 54].

Electric vehicle charging technology goes through quick advancements to satisfy increasing EV market requirements and to solve speed and convenience-related issues as well as grid integration needs. The EV charging industry will experience expansion because of hardware and software usage in charging stations while focusing on developments to enhance both charging efficiency and grid stability. Electric vehicle production has increased significantly throughout the last decade because people have become more aware of environmental issues, stricter emission standards, and better battery technology has emerged. Figure 2 illustrates different types of EV infrastructure. The diagram outlines the main parts of today's EV infrastructure, which include numerous charging methods, smart grid connections, helping with

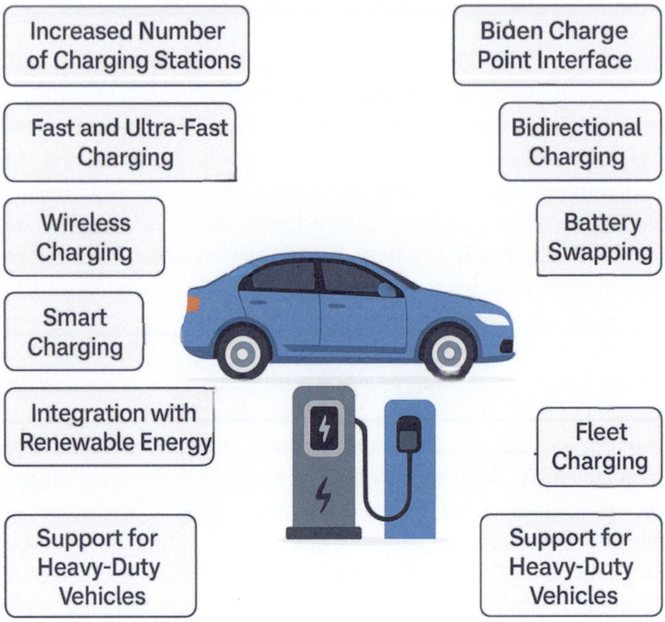

Fig. 2 EV infrastructure

renewable energy, and services suitable for a range of vehicles and fleets. Every single element works with the EV, making sure charging is effective and thorough.

3 Control Topologies in EV Charging Systems

The charging station received a precise design to integrate effortlessly with electrical networks as well as maintain compatibility with future technological upgrades. The development of charging technologies presents several options for battery exchange systems, dynamic charging methods, and inductive charging solutions, which enhance availability and safety performance and speed up the charging process. A combination of different control topologies exists in EV charging stations to execute power management and guarantee grid stability. Different levels of "uncontrolled" and "smart" charging methods make up the range of topologies that exist in the EV charging domain.

This analysis depicts different control methods for electric vehicles.

1. Uncontrolled Charging: The uncontrolled charging method represents the most basic approach, which enables electric vehicles to pull power straight from the electrical grid without any management [36]. Easy implementation of this method

results in power grid instability, particularly when electric vehicles make up a large portion of the system. The installation of further charging stations increases EV owner comfort while shortening queue times when charging [36].

2. Controlled Charging: This method, named Controlled Charging (V1G), allows management of EV charging speeds and timing according to electrical grid situations and power price trends [36]. Such changes include pricing models that encourage vehicle charging when energy demand is lowest. WINSmartEV serves as a leading system that integrates intelligent control elements across diverse levels to implement charge scheduling systems, which prevent network congestion [55].
3. Vehicle-to-Grid: Users equipped with V2G technology benefit from dual electric vehicle operations by pulling power from the grid, combined with sending stored power to the grid [56]. Such systems need complicated control systems and two-way power flow capacity [36]. The implementation of V2G requires Ultra-Fast DC Chargers to achieve this functionality [56].
4. Master–Slave Control: The topology design of Master–Slave Control features a central controller that oversees more than one charging point. A round-robin algorithm operates as a basic system to activate only a single channel for charging one PEV at a time within Level 1 charging stations [55]. The central controller maintains capabilities of performance updates through having flexibility and extensibility features to adopt newly developed algorithms [55].
5. Internal Model Control: Internal Model Control represents a sophisticated technique that finds implementation in EV chargers to boost performance along with dual control of AC and DC fast charging operations [40]. The implementation of harmonic generation and control complexity, along with cost, remains an ongoing challenge [40].

Various optimization methods exist for enhancing the operating performance of control topologies, both in [15, 57, 58]. Searching heuristically and applying mathematical programming helps optimization systems work effectively and correctly [15]. The development of improved control methods can lead EV charging facilities to advance their speed of operation and convenience while integrating better into the power grid. Figure 3 EV Topology: It outlines EV charging, showing how on-board and off-board chargers connect, AC-DC and DC-DC conversion take place, and many types of power converters are involved. Looking at power, charge both ways, and vehicle-to-grid abilities, we see how each factor fits the EV to make charging flexible and manageable.

4 Conclusion

This Chapter concludes that the next generation of EV charging infrastructure must be intelligent, resilient, and seamlessly integrated with renewable energy systems and smart grids. High-power charging technologies such as ultra-fast DC and dynamic

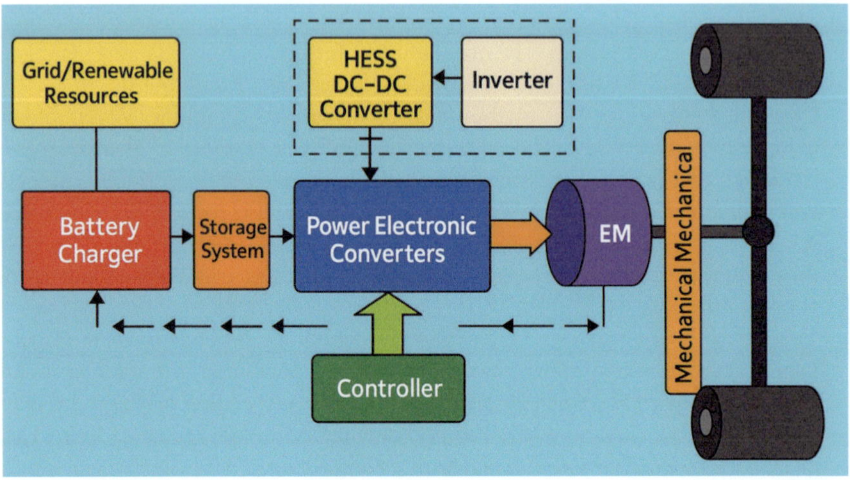

Fig. 3 EV topology

inductive charging are essential to address range anxiety and reduce vehicle downtime. Smart charging strategies—enabled by AI, machine learning, and predictive analytics—are necessary for synchronizing vehicle charging with grid capacity and renewable energy availability. Additionally, vehicle-to-grid (V2G) systems must be supported by bi-directional converters and robust control mechanisms to facilitate energy arbitrage, grid support, and demand-side management. Advanced control topologies, including internal model control, centralized and decentralized algorithms, and adaptive power flow coordination, are required to maintain grid integrity under high EV penetration. Governmental incentives, standardized protocols (e.g., Plug & Charge), and strategic deployment of infrastructure are pivotal for ensuring equitable access and user convenience. Future research should focus on the economic viability of wireless and dynamic charging models, real-time optimization in heterogeneous traffic environments, and the co-design of mobility and energy systems. Through multi-disciplinary collaboration, the full potential of EV adoption as a sustainable, grid-interactive technology can be realized.

References

1. Yilmaz M, Krein PT (2013) Review of battery charger topologies, charging power levels, and infrastructure for plug-in electric and hybrid vehicles. IEEE Trans Power Electron 28(5):2151–2169. https://doi.org/10.1109/TPEL.2012.2212917
2. Khaligh A, Dusmez S (2012) Comprehensive topological analysis of conductive and inductive charging solutions for plug-in electric vehicles. IEEE Trans Veh Technol 61(8):3475–3489. https://doi.org/10.1109/TVT.2012.2198676
3. Wirges A, Linder T, Kessler W (2012) Modelling the development of a regional charging infrastructure for electric vehicles in time and space. Eur J Transp Infrastruct Res 12(4):391–416

4. Habib S, Kamran M, Rashid U (2015) Impact analysis of vehicle-to-grid technology and charging strategies of electric vehicles on distribution networks—a review. J Power Sources 277:205–214. https://doi.org/10.1016/j.jpowsour.2014.12.020
5. Zhang Y, Yu Y, Yu X (2018) Electric vehicle charging load forecasting with long short-term memory network. IEEE Smart Grid Comm 1–6. https://doi.org/10.1109/SmartGridComm.2018.8587494
6. Jacobson MZ et al (2017) 100% clean and renewable wind, water, and sunlight all-sector energy roadmaps for 139 countries of the world. Joule 1(1):108–121. https://doi.org/10.1016/j.joule.2017.07.005
7. Emadi A, Rajashekara K, Williamson SS, Lukic SM (2007) Topological overview of automotive power electronics. IEEE Trans Industr Electron 54(2):557–570. https://doi.org/10.1109/TIE.2007.891632
8. Lukic S, Pantic Z (2013) Cutting the cord: static and dynamic inductive wireless charging of electric vehicles. IEEE Electrification Mag 1(1):57–64. https://doi.org/10.1109/MELE.2013.2277987
9. Slootweg JG, Kling WL (2002) The impact of distributed generation on power system transient stability. Electr Power Syst Res 62(1):1–9. https://doi.org/10.1016/S0378-7796(02)00037-9
10. Dwivedi S, Khopkar AS, Tailor H (2025) Optimal location of surge arrester for protection of 220kV transmission line against lightning over voltages using EMTP-RV software. Eng Technol J 10(1). https://doi.org/10.47191/etj/v10i01.14
11. Hans N (2020) Trends in electric vehicle (EV) charging and key technology developments. Int J Eng Res 9. https://doi.org/10.17577/ijertv9is090042
12. Khan W, Ahmad A, Ahmad F, Alam MS (2018) A comprehensive review of fast charging infrastructure for electric vehicles. Smart Sci 1. https://doi.org/10.1080/23080477.2018.1437323
13. Deb S, Tammi K, Kalita K, Mahanta P (2018) Impact of electric vehicle charging station load on distribution network. Energies 11(1):178. https://doi.org/10.3390/en11010178
14. Geng S et al (2023) Challenges and opportunities for fast-charging batteries. J Phys Chem C 127(31):15021. https://doi.org/10.1021/acs.jpcc.3c01927
15. Si R et al (2025) Hybrid optimization-based sequential placement of des in unbalanced active distribution networks considering multi-scenario operation. Energies 18(3):474. https://doi.org/10.3390/en18030474
16. Fernández E, Albizu I, Bedialauneta MT, Mazón AJ, Asano PTL (2015) Review of dynamic line rating systems for wind power integration. Renew Sustain Energy Rev 53:80. https://doi.org/10.1016/j.rser.2015.07.149
17. Aznarte JL, Siebert N (2016) Dynamic line rating using numerical weather predictions and machine learning: a case study. IEEE Trans Power Deliv 32(1):335. https://doi.org/10.1109/tpwrd.2016.2543818
18. Yong JY, Ramachandaramurthy VK, Tan KM, Mithulananthan N (2015) A review on the state-of-the-art technologies of electric vehicle, its impacts and prospects. Renew Sustain Energy Rev 49:365. https://doi.org/10.1016/j.rser.2015.04.130
19. Michiorri A et al (2015) Forecasting for dynamic line rating. Renew Sustain Energy Rev 52:1713. https://doi.org/10.1016/j.rser.2015.07.134
20. Next generation dynamic line rating provides strong economic benefits. Accessed: 09 Apr 2025. Available: https://electricenergyonline.com/energy/magazine/970/article/Next-Generation-Dynamic-Line-Rating-Provides-Strong-Economic-Benefits.htm
21. Teng F, Dupin R, Michiorri A, Kariniotakis G, Chen Y, Štrbac G (2017) Understanding the benefits of dynamic line rating under multiple sources of uncertainty. IEEE Trans Power Syst 33(3):3306. https://doi.org/10.1109/tpwrs.2017.2786470
22. An-nizha N, Sofyan S, Sultan AR, Bachtiar MI, Umar M, Arni S (2024) Genetic algorithm-based optimization of distributed generator placement in distribution networks to minimize power losses. J Teknol Elekterika 21(2):157. https://doi.org/10.31963/elekterika.v21i2.5117
23. Kumar M, Sharma D (2025) Correction: energy management in microgrids with uncertainty in EV and renewable sources. Iran J Sci Technol Trans Electr Eng. https://doi.org/10.1007/s40998-025-00794-6

24. Nguyen DM, Kishk MA, Alouini M (2024) Dynamic charging as a complementary approach in modern EV charging infrastructure. Sci Rep 14(1). https://doi.org/10.1038/s41598-024-55863-3
25. Shibl MM, Ismail L, Massoud A (2020) Machine learning-based management of electric vehicles charging: towards highly-dispersed fast chargers. Energies 13(20):5429. https://doi.org/10.3390/en13205429
26. Rabih M, Takruri M, Al-Hattab M, Alnuaimi AA, Thaleth MRB (2024) Wireless charging for electric vehicles: a survey and comprehensive guide. World Electr Veh J 15(3):118. https://doi.org/10.3390/wevj15030118
27. Liu H, Wang DZW (2017) Locating multiple types of charging facilities for battery electric vehicles. Transp Res Part B Methodol 103:30. https://doi.org/10.1016/j.trb.2017.01.005
28. Rahulkumar J et al (2022) An empirical survey on wireless inductive power pad and resonant magnetic field coupling for in-motion EV charging system. IEEE Access 11:4660. https://doi.org/10.1109/access.2022.3232852
29. Hata K, Imura T, Hori Y (2015) Dynamic wireless power transfer system for electric vehicle to simplify ground facilities—power control based on vehicle-side information. World Electr Veh J 7(4):558. https://doi.org/10.3390/wevj7040558
30. Panchal C, Stegen S, Lu J (2018) Review of static and dynamic wireless electric vehicle charging system. Eng Sci Technol Int J 21(5):922. https://doi.org/10.1016/j.jestch.2018.06.015
31. Karakitsios I, Karfopoulos E, Hatziargyriou N (2014) Static and dynamic fast inductive charging: the fast in charge project concept. https://doi.org/10.1049/cp.2014.1663
32. Yan F, XF (2016) Analysis on influencing factors of EV charging station planning based on AHP. In: IOP conference series earth and environmental science. IOP Publishing, p 12054. https://doi.org/10.1088/1755-1315/40/1/012054
33. Muttaqi KM, Isac E, Mandal A, Sutanto D, Akter S (2023) Fast and random charging of electric vehicles and its impacts: state-of-the-art technologies and case studies. Electr Power Syst Res 226:109899. https://doi.org/10.1016/j.epsr.2023.109899
34. Seth AK, Singh M (2021) Unified adaptive neuro-fuzzy inference system control for OFF board electric vehicle charger. Int J Electr Power Energy Syst 130:106896. https://doi.org/10.1016/j.ijepes.2021.106896
35. Collin R, Yu M, Yokochi A, Enjeti P, von Jouanne A (2019) Advanced electric vehicle fast-charging technologies. Energies 12(10):1839. https://doi.org/10.3390/en12101839
36. Qing K, Fowler M, Entchev E, Ribberink H, McCallum R (2018) The role of charging infrastructure in electric vehicle implementation within smart grids. Energies 11(12):3362. https://doi.org/10.3390/en11123362
37. Ayoade IA, Longe OM (2024) A comprehensive review on smart electromobility charging infrastructure. World Electr Veh J 15(7):286. https://doi.org/10.3390/wevj15070286
38. Bons P, Buatois A, Schuring F, Geerts F, van den Hoed R (2021) Flexible charging of electric vehicles: results of a large-scale smart charging demonstration. World Electr Veh J 12(2):82. https://doi.org/10.3390/wevj12020082
39. Ilahi T et al (2024) Design analysis of high-power level 4 smart charging infrastructure using next-generation power devices for EVs and heavy duty EVs. World Electr Veh J 15(2):66. https://doi.org/10.3390/wevj15020066
40. Al-Ogaili AS, Aris I, Ramasamy AK, Hashim TJT, Marsadek MB, Sabry AH (2020) Integrating internal model controller (IMC) into electric vehicle charger of multiple charging mode: dc and ac fast charging. Appl Sci 10(12):4179. https://doi.org/10.3390/app10124179
41. Deilami S, Muyeen SM (2020) An insight into practical solutions for electric vehicle charging in smart grid. Energies 13(7):1545. https://doi.org/10.3390/en13071545
42. Chakraborty P et al (2022) Addressing the range anxiety of battery electric vehicles with charging en route. Sci Rep 12(1). https://doi.org/10.1038/s41598-022-08942-2
43. Zou T, Khaloei M, MacKenzie D (2020) Effects of charging infrastructure characteristics on electric vehicle preferences of new and used car buyers in the United States. Transp Res Rec J Transp Res Board 2674(12):165. https://doi.org/10.1177/0361198120952792

44. Jebril IH et al (2021) User satisfaction of electric-vehicles about charging stations (Home, Outdoor, and Workplace). Türk Bilgi Ve Mat Eğitimi Derg 12(3):3589. https://doi.org/10.17762/turcomat.v12i3.1637
45. Ren L, Liao W, Chen J (2024) Systematic design and implementation method of battery-energy comprehensive management platform in charging and swapping scenarios. Energies 17(5):1237. https://doi.org/10.3390/en17051237
46. Ajao Q, Sadeeq L (2023) An approximate feasibility assessment of electric vehicles adoption in Nigeria: Forecast 2030. arXiv (Cornell University). https://doi.org/10.48550/arxiv.2305.17844
47. Srihari G, Naidu RS, Falkowski-Gilski P, Parameshachari BD, Varma PRK (2024) Integration of electric vehicle into smart grid: a meta heuristic algorithm for energy management between V2G and G2V. Front Energy Res 12. https://doi.org/10.3389/fenrg.2024.1357863
48. Tan X, Leon-Garcia A (2018) Autonomous mobility and energy service management in future smart cities: an overview. arXiv (Cornell University). https://doi.org/10.48550/arxiv.1810.04576
49. Mishra S et al (2021) A comprehensive review on developments in electric vehicle charging station infrastructure and present scenario of India. Sustainability 13(4):2396. https://doi.org/10.3390/su13042396
50. Sun X, Li Z, Wang X, Li C (2019) Technology development of electric vehicles: a review. Energies 13(1):90. https://doi.org/10.3390/en13010090
51. Mastoi MS et al (2022) An in-depth analysis of electric vehicle charging station infrastructure, policy implications, and future trends. Energy Rep 8:11504. https://doi.org/10.1016/j.egyr.2022.09.011
52. Amilia N, Palinrungi Z, Vanany I, Arief M (2022) Designing an optimized electric vehicle charging station infrastructure for urban area: a case study From Indonesia. arXiv (Cornell University). https://doi.org/10.48550/arxiv.2209.03448
53. Alexeenko P, Bitar E (2023) Achieving reliable coordination of residential plug-in electric vehicle charging: a pilot study. Transp Res Part D: Transp Environ 118:103658. https://doi.org/10.1016/j.trd.2023.103658
54. Kundu S, Singh M, Giri AK et al (2024) Robust and fast control approach for islanded microgrid system and EV charging station applications. Electr Eng. https://doi.org/10.1007/s00202-024-02291-1
55. Chung C-Y, Chynoweth J, Chu C, Gadh R (2014) Master-slave control scheme in electric vehicle smart charging infrastructure. Sci World J 2014:1. https://doi.org/10.1155/2014/462312
56. Krishna B, Anusha D, Karthikeyan V (2020) Ultra-fast DC charger with improved power quality and optimal algorithmic approach to enable V2G and G2V. J Circuits Syst Comput 29(12):2050197. https://doi.org/10.1142/s0218126620501972
57. Kumar M, Sharma D (2025) Energy management in microgrids with uncertainty in EV and renewable sources. Iran J Sci Technol Trans Electr Eng 49:639–661. https://doi.org/10.1007/s40998-024-00786-y
58. Kundu S, Giri AK, Kadiyan S (2025) An adaptive mixed-step size normalized least means fourth control approach for stand-alone power generation system considering dynamic conditions. IEEE J Emerg Sel Top Power Electron 13(2):2422–2429. https://doi.org/10.1109/JESTPE.2024.3516878
59. Singh RK, Rajesh MR (2020) Smart grid technologies for electric vehicle integration: a review. Renew Sustain Energy Rev 120:109618. https://doi.org/10.1016/j.rser.2019.109618

MIX
Papier aus verantwortungsvollen Quellen
Paper from responsible sources
FSC® C105338

If you have any concerns about our products,
you can contact us on
ProductSafety@springernature.com

In case Publisher is established outside the EU,
the EU authorized representative is:
Springer Nature Customer Service Center GmbH
Europaplatz 3, 69115 Heidelberg, Germany

Printed by Libri Plureos GmbH
in Hamburg, Germany